Seagrass Ecology

Seagrasses occur widely in coastal zones throughout the world, in the part of the marine habitat that is most heavily influenced by humans. Decisions about coastal management therefore often involve seagrasses, but despite a growing awareness over the past few decades of the importance of these plants (for example, with respect to the regulation of biogeochemical cycles), a full appreciation of the role of seagrasses in coastal ecosystems has yet to be reached This book provides an entry point for those wishing to learn about the ecology of this fascinating group of plants, and gives a broad overview of the present state of knowledge, including recent progress in research and current research foci, complemented by extensive literature references to guide the reader to more detailed studies. As such it will be valuable to students of marine biology wishing to specialize in this area and also to established researchers wanting to enter the field. In addition, it will provide an excellent source of information for those involved in the management and conservation of coastal areas that harbour seagrasses.

MARTEN HEMMINGA is Director of the Dutch nature conservation agency Stichting Het Zeeuwse Landschap, a position he took up in 1999 after nine years as Head of the Department of Littoral Vegetation at the Netherlands Institute of Ecology's Centre for Estuarine and coastal Ecology, Yerseke. He has also held a part-time chair in Estuarine Ecology at the University of Nijmegen. His research interests focus on carbon and nutrient dynamics in seagrass systems, salt marshes and mangroves.

CARLOS DUARTE is a Research Professor at the Spanish Research Council's Instituto Mediterráneo de Estudios Avanzados at Palma de Mallorca, where his research focuses on seagrass ecosystems, the ecology of coastal ecosystems and the comparative ecology of marine ecosystems.

Seagrass Ecology

Marten A. Hemminga, 1954–
*Netherlands Institute of Ecology, Yerseke, and
Stichting Het Zeeuwse Landschap, Heinkenszand*

Carlos M. Duarte
*Instituto Mediterráneo de Estudios Avanzados,
Palma de Mallorca*

CAMBRIDGE
UNIVERSITY PRESS

PUBLISHED BY THE PRESS SYNDICATE OF THE UNIVERSITY OF CAMBRIDGE
The Pitt Building, Trumpington Street, Cambridge, United Kingdom

CAMBRIDGE UNIVERSITY PRESS
The Edinburgh Building, Cambridge CB2 2RU, UK
40 West 20th Street, New York, NY 10011–4211, USA
10 Stamford Road, Oakleigh, VIC 3166, Australia
Ruiz de Alarcón 13, 28014 Madrid, Spain
Dock House, The Waterfront, Cape Town 8001, South Africa

http://www.cambridge.org

© Cambridge University Press 2000

First published 2000

Printed in the United Kingdom at the University Press, Cambridge

Typeface Times 10/13pt   *System* 3B2   [CE]

A catalogue record for this book is available from the British Library

Library of Congress Cataloguing in Publication data

Hemminga, Marten, 1954–
Seagrass ecology / Marten Hemminga, Carlos Duarte.
    p. cm.
ISBN 0 521 66184 6
1. Seagrasses – Ecology. I. Duarte, Carlos, 1960–. II. Title.
QK495.A14 H46   2000
577.69′4–dc21   00–037917

ISBN 0 521 66184 6 hardback

Contents

Preface

Only a few decades ago, seagrass ecology was a virtually non-existent field within marine ecology. In the past 30 years or so, this situation has drastically changed. As a recent analysis of published papers by one of us (CMD; *Aquatic Botany* 65: 7–20) indicates, current publication rates on seagrass ecology in the international scientific literature are at a level of approximately 100 papers per year, and these rates are still increasing. The growing awareness of the role that seagrasses play in ecology, and in the regulation of the biogeochemical cycles of the coastal zones world-wide, undoubtedly does much to stimulate this ongoing expansion of research efforts. The increase in the number of studies also implies that the community of researchers that enters the field of seagrass studies is growing. Besides the scientists, there is also an expanding community of professionals working in governmental and non-governmental organizations in countries all over the world, which in the context of coastal management or conservation issues are actively involved in seagrass matters.

This book is intended to provide an introduction to the field of seagrass ecology. In the first place it is designed for students and for scientists who enter an area as yet unfamiliar to them. In the second place, we hope that the book will also serve as a source of information for those involved in management and conservation of coastal areas that harbour seagrasses. In writing this book, we made several choices that determined the design of the book and its chapters. We wanted it to provide the reader with a broad picture of the array of topics that are currently studied in the field of seagrass ecology. Furthermore, we wanted the content to present the state-of-the-art in the various research directions. Finally, we also wanted the book to facilitate more thorough study of the different topics that currently receive major attention, by adding extensive literature references to each chapter. Moreover, these aims had to be realized in a book with a modest page number, requiring a concise treatment of the different topics. Our choice, therefore, has been to concentrate on the literature that has been published in the 1990s wherever the topic allowed this, without neglecting the foundation that was laid in earlier decades. We feel that this pragmatic choice is justified, particularly so because much of the older literature is adequately covered in earlier books that appeared in the 1970s and 1980s.

The book starts with an introductory chapter on taxonomy and distribution, which outlines the taxonomic division and global bio-geography of the seagrass flora, and discusses the occurrence of sea-grasses in relation to their primary habitat requirements. The second chapter deals with seagrass architectural features. At first sight this may seem odd in an ecological book, yet the architecture of seagrasses, the fact that they are all clonal, rhizomatous plants, strongly determines their performance in the marine environment. The clonal design of seagrasses and the existence of large differences in growth rates between species, which are essentially governed by scaling laws that have been elucidated in the past decade, are of eminent importance in explaining the growth patterns of different species. The third chapter on population and community dynamics of seagrasses reflects this. This chapter focuses on the fluctuations in abundance at different levels of complexity, from the level of individual modules such as leaves, to that of entire meadows. The fourth chapter focuses on major environmental resources required for seagrass growth, i.e. light, inorganic carbon and nutrients. Many new studies have been carried out in this field in the past decade, together greatly contributing to insight in the different type of growth limitations that seagrasses face in the marine realm. The fifth chapter deals with the elemental dynamics in seagrass systems. This is a broad topic: a variety of processes contributes to the fluxes of matter in seagrass systems, and the chapter addresses such diverse processes as herbivory, burial and advective transports. Benthic mineralization processes are emphasized, as much progress has been made in this field, allowing comparisons between processes and evaluation of the balance between primary production and mineralization at the system level. No single subject in the field of seagrass studies, however, has received more attention than that of the fauna that is associated with seagrasses. This topic is addressed in chapter six. First, we discuss the general abundance and diversity of fauna in seagrass systems. Fishes, crustaceans, molluscs and the large vertebrate grazers turtles and dugongs are treated separately thereafter. The significance of seagrass meadows as a habitat and a foraging area to these animals is a recurrent theme in each of these sections. Seagrasses abound in coastal areas, the part of the marine environment that is most heavily influenced by humans. It is thus not surprising that the numerous declines of seagrass populations reported all over the world for the major part have an anthropogenic cause. The impact of human activities on the vitality and persistence of seagrasses is described in the last, seventh chapter.

Through discussions, suggestions, materials and review of different chapters, many colleagues indirectly contributed to this book. In particular we thank the following colleagues: Quique Ballesteros, Sven Beer, Claire Billot, Erik Boschker, Robert Coles, Mike Durako, Mike Fortes, Jim Fourqurean, Kees den Hartog, Marieke van Katwijk, Jud Kenworthy, Núria Marbà, Karen McGlathery, Jack Middelburg, Bob Orth, Fred Short, Japar Sidik Bujang, Jorge Terrados, Jan Vermaat, Diana Walker and Michelle Waycott. We are, of course, the only individuals responsible for any omissions or errors this book may contain. CMD warmly thanks Kaj Sand-Jensen and Mike Fortes for guidance and inspiration in the last decade. We dedicate this book to our families, whose patience and support during the time it was produced made it possible.

Marten Hemminga (Yerseke)
Carlos Duarte (Palma de Mallorca)

December 1999

1

Taxonomy and distribution

1.1 Introduction

Seagrasses comprise < 0.02% of the angiosperm flora, representing a surprisingly small number of species (about 50, Table 1.1) compared with any other group of marine organisms. The limited species membership of the seagrass flora has directed some (still limited) efforts to the study of their origin and their evolution in an attempt to account for this phenomenon. A second path of research has tried to find clues for the paucity of species by studying the stress factors constraining angiosperm life in the sea. This second approach has driven much effort towards the analysis of seagrass distribution and the definition of the habitat requirements of seagrasses. The attention these issues have received extends beyond scholarly concerns, for seagrasses are, despite their limited diversity, important contributors to coastal marine ecosystems, both locally and at the global scale. In this chapter we shall provide an overview of the origin, evolution and present diversity of extant seagrasses, and describe their present distribution and the basic requirements that delimit their possible habitats. The definition of how seagrass distribution is regulated leads, in turn, to the assessment of their global extent and, from this, to the evaluation of the role seagrasses play on the global ocean ecosystem.

1.2 The seagrass flora

Seagrasses are generally assigned to two families, Potamogetonaceae and Hydrocharitaceae, encompassing 12 genera of angiosperms containing about 50 species (Table 1.1). Three of the genera, *Halophila, Zostera* and *Posidonia*, which may have evolved from lineages that appeared relatively

1

Table 1.1. *List of seagrass species indicating their membership to the different seagrass floras*

Species	Biogeographic membership
Amphibolis antarctica	S. Australian flora
Amphibolis griffithii	S. Australian flora
Cymodocea angustata	Indo-Pacific flora
Cymodocea nodosa	Mediterranean flora
Cymodocea rotundata	Indo-Pacific flora
Cymodocea serrulata	Indo-Pacific flora
Enhalus acoroides	Indo-Pacific flora
Halodule pinifolia	Indo-Pacific flora
Halodule uninervis	Indo-Pacific flora
Halodule wrightii	Caribbean flora
Halophila baillonis	Caribbean flora
Halophila beccarii	Indo-Pacific flora
Halophila capricornii	Indo-Pacific flora
Halophila decipiens	Caribbean and Indo-Pacific floras
Halophila engelmannii	Caribbean flora
Halophila hawaiiana	Indo-Pacific flora
Halophila ovalis	Indo-Pacific flora
Halophila ovata	Indo-Pacific flora
Halophila spinulosa	Indo-Pacific flora
Halophila stipulacea	Indo-Pacific flora
Heterozostera tasmanica	S. Australian flora
Phyllospadix iwatensis	Temperate W. Pacific flora
Phyllospadix japonicus	Temperate W. Pacific flora
Phyllospadix scouleri	Temperate E. Pacific flora
Phyllospadix serrulatus	Temperate E. Pacific flora
Phyllospadix torreyi	Temperate E. Pacific flora
Posidonia angustifolia	S. Australian flora
Posidonia australis	S. Australian flora
Posidonia coriacea	S. Australian flora
Posidonia denhartogii	S. Australian flora
Posidonia kirkmanii	S. Australian flora
Posidonia oceanica	Mediterranean flora
Posidonia ostenfeldii	S. Australian flora
Posidonia robertsoniae	S. Australian flora
Posidonia sinuosa	S. Australian flora
Syringodium filiforme	Caribbean flora
Syringodium isoetifolium	Indo-Pacific flora
Thalassia hemprichii	Indo-Pacific flora
Thalassia testudinum	Caribbean flora
Thalassodendron ciliatum	Indo-Pacific flora
Thalassodendron pachyrhizum	S. Australian flora
Zostera asiatica	Temperate W. Pacific flora
Zostera capensis	S. Atlantic flora
Zostera capricorni	S. Australian flora
Zostera caulescens	Temperate W. Pacific flora
Zostera japonica	Temperate W. Pacific flora

Table 1.1. (*cont.*)

Species	Biogeographic membership
Zostera marina	N. Atlantic, Mediterranean, W. and E. Pacific floras
Zostera mucronata	S. Australian flora
Zostera muelleri	S. Australian flora
Zostera noltii	N. Atlantic and Mediterranean floras
Zostera novazelandica	New Zealand flora

Note: A number of additional species, particularly within the genus *Halophila*, have been described, but on-going examinations are suggesting that the number of species be revised downwards (Waycott, 1999).
Source: After Phillips & Meñez, 1988; Kirkman & Walker, 1989.

early in seagrass evolution (Den Hartog, 1970), comprise most (55%) of the species, while *Enhalus*, the most recent seagrass genus, is represented by a single species (*Enhalus acoroides*, Table 1.1).

Despite the limited size of the seagrass flora, there is no generally accepted number of species, and an examination of the literature indicates that there is a 20% uncertainty about what the total number of seagrass species is (e.g. Waycott, 1999). This uncertainty results largely from disagreement about the number of species in the three genera with the largest number of species (i.e. *Halophila*, *Zostera* and *Posidonia*). This disagreement largely reflects differences in the criteria used to classify seagrasses. Current identification keys are based, at the species level, on differences in the shape of the margins and venation of leaves as diagnostic criteria (Den Hartog, 1970; Phillips & Meñez, 1988). More conclusive criteria, such as the structure of reproductive organs, are of limited use for identification, since they are highly reduced in most seagrass species and they are difficult to find in some species (see Chapter 3). The extent of variation in the characters used for taxonomic purposes is not well resolved, and there is, in general, a considerable uncertainty derived from plasticity within seagrass species. It is, therefore, hardly surprising that some scientists using morphological and anatomical characters may be inclined to include a larger number of species, whereas those using the more parsimonious techniques of molecular taxonomy question even the basis for separation of some of the genera. Les *et al.* (1997) have recently reported, on the basis of molecular phylogenetic analyses, that the genetic distance between the *Heterozostera* and *Zostera* species is similar to that between species of *Zostera*, recommending that

these two genera be merged into a single genus. Hence, there is a great
need to quantify morphological and genetic plasticity of seagrass species
and conduct genetic studies across large scales to resolve present uncertainties on the taxonomic status of seagrasses. Even differences in key
life-history traits, such as annual or perennial life cycles, are not enough
to determine genetic differences, as demonstrated by the genetic identity
of annual and perennial forms of *Zostera marina* (Gagnon *et al.*, 1980).

1.3 The origin and evolution of seagrasses

The small size of the seagrass flora might be considered as an indicator of
a recent origin, but the fossil record and indirect evidence (e.g. fossils
from associated fauna) fail to support this suggestion and point, in
contrast, to an early origin in the evolution of angiosperms. Existing
evidence indicates that angiosperms colonized the marine environment
about 100 million years ago (Den Hartog, 1970), which indicates a
relatively early appearance of seagrasses in angiosperm evolution, as
compared with the accepted origin of angiosperms about 400 million
years ago (Raven, 1977).

Candidate seagrass ancestors are coastal plants and freshwater hydrophytes (Larkum & Den Hartog, 1989). The hypothesis of coastal plants
(i.e. marsh plants or mangroves) as ancestors of seagrasses has been
defended on the grounds that some seagrass genera have lignified stems,
compared with the herbaceous stems of all hydrophytes, and that two
seagrass genera (*Amphibolis* and *Thalassodendron*) are viviparous, as is
the case in some mangrove taxa (cited in Larkum & Den Hartog, 1989).
The rationale behind the hypothesis that seagrasses originate from
hydrophytes is that the latter show many of the adaptations that have
been postulated to be necessary for angiosperm life in the marine
environment (e.g. basal meristems, extensive lacunar systems; cf. Arber,
1920). Although it now seems that all angiosperms, including terrestrial
ones, derive from early hydrophytes, there is no clear evidence that the
specific hydrophytes hypothesized to be the seagrass ancestors evolved
prior to them. Furthermore, one of the oldest fossil seagrasses, *Thalassocharis*, lacked a lacunar system in its stems (Larkum & Den Hartog,
1989). The seagrass fossil record is very poor, the oldest Cretaceous
specimens belonging to, at most, three genera, including *Thalassocharis*
and *Posidonia* (Den Hartog, 1970). There is evidence that most modern
seagrass genera were already established in the late Eocene, about 40
million years before present (Larkum & Den Hartog, 1989). The genera

Phyllospadix and *Enhalus* seem to have appeared long after other sea-grass genera were established (Larkum & Den Hartog, 1989). In the case of *Enhalus acoroides*, the only seagrass species lacking hydrophilous pollination (see Chapter 3), it has been suggested that this species evolved from a freshwater ancestor, the genus *Vallisneria* (Larkum & Den Hartog, 1989), which *Enhalus* resembles both in form and pollination mode.

Given the paucity of fossils, a robust phylogeny of seagrasses must be based on molecular analyses of DNA sequences, which are only now becoming available (e.g. Les *et al.*, 1997). Chemotaxonomic evidence, such as that based on the forms of sugars present, has yielded results consistent with the hypothesis of a polyphyletic origin of seagrasses (e.g. Drew, 1983). More recently, Les *et al.* (1997) examined gene sequences of seagrass chloroplasts to provide support for the hypothesis that there are multiple origins of seagrasses. These origins appear to involve freshwater ancestors for the Hydrocharitaceae, a saltmarsh or aquatic ancestor for the Zosteraceae, and an ancestor of *Ruppia* for the third group, to which it is closely linked (Les *et al.*, 1997).

The evolutionary history of seagrasses has involved the acquisition of key adaptations necessary for successful colonization of the marine habitat: (1) blade or subulate leaves with sheaths, adapted to high-energy environments; (2) hydrophilous pollination, allowing submarine pollination (except for the genus *Enhalus*, cf. Chapter 3); and (3) extensive lacunar systems allowing the internal gas flow needed to maintain the oxygen supply required by their belowground structures in anoxic sediments.

The path of seagrass evolution is relatively unclear, but speciation certainly has been remarkably conservative. Although some extinctions have been documented in the fossil record, examination of the evidence present in the literature suggests that the total number of species may never have significantly exceeded the present number. The possible reasons for the paucity of species have been the subject of much discussion (e.g. Van der Hage, 1996; Ackerman, 1998). Most arguments converge to the notion that the low rate of sexual reproduction and the low-range dispersion associated with hydrophilous pollination of most seagrasses have restricted their gene flow to a small neighbourhood, reducing seagrass genetic diversity compared with their terrestrial counterparts. Yet these explanations are not fully supported by the growing estimates of genetic diversity, which show that this can be substantial in some meadows (section 1.5).

1.4 Seagrass biogeography and distribution of seagrass floras

Seagrasses occur in all coastal areas of the world, except along Antarctic shores, probably because ice scouring, which greatly damages seagrasses (Robertson & Mann, 1984), renders such areas unsuitable for seagrass life. Seagrass species are often separated into tropical and temperate genera. The former are considered to comprise seven of the genera, while the latter comprise the remaining five genera. There are, however, too many exceptions to this division for it to be used as a general classification. For instance, *Cymodocea nodosa*, considered a tropical genus, is widespread along the temperate shores of the Mediterranean Sea, the Atlantic shores of south Portugal and North-West Africa (Den Hartog, 1970), whereas species of *Zostera*, considered a temperate genus, are also encountered along tropical coasts, such as the populations of *Zostera japonica* along the coast of Vietnam. It is probably more accurate to discriminate between nine different seagrass floras (Table 1.1, Fig. 1.1), which can be summarized, following parsimonic criteria, into:

1 A temperate North Atlantic flora.
2 A temperate East Pacific flora.
3 A temperate West Pacific flora.
4 A temperate South Atlantic flora.
5 A Mediterranean flora.
6 A Caribbean flora.
7 An Indo-Pacific flora.
8 A South Australia flora.
9 A New Zealand flora.

These floras are not always separated by clear boundaries, and their biogeographic ranges include, therefore, contact zones, such as the subtropical Atlantic zone, which contains *Zostera noltii* and *Cymodocea nodosa* on its eastern shore, and *Halodule wrightii* on its western and eastern shores. Although seagrasses are present along the coasts of South America and West Africa, their abundance on both sides of the South Atlantic appears to be very small, This, however, may only reflect the paucity of studies in these areas.

The distribution of seagrass floras outlined above contains several salient features. Floras containing 'twin species', a term used to refer to congeneric species present in different floras, or assemblages of them, are evident. The occurrence of a twin assemblage of *Thalassia–Halodule–Syringodium* in the Caribbean and the Indo-Pacific region (Table 1.1,

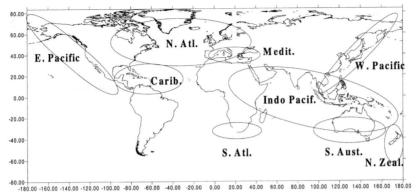

Fig. 1.1. The distribution of seagrass floras in the world: temperate North Atlantic flora, temperate East Pacific flora, temperate West Pacific flora, temperate South Atlantic flora, Mediterranean flora, Caribbean flora, Indo-Pacific flora, South Australian flora, and New Zealand flora.

Fig. 1.1) has received considerable attention. Twin, congeneric floras that include *Zostera* and *Phyllospadix* species are also present along both shores of the temperate North Pacific (Table 1.1, Fig. 1.1). Twin species imply a fragmentation of the distribution of the genus, with all *Posidonia* species (except the Mediterranean endemic species *Posidonia oceanica*), restricted to South Australian waters, and one of the two *Thalasso-dendron* species being present in the Indo-Pacific area, and the other in West Australia. It is possible that the recognition of some of the congeneric species present in twin assemblages as different species is solely due to their geographic separation, a possibility that must be ruled out with the use of robust molecular taxonomical techniques.

The attempts to account for the occurrence of twin species and fragmented distribution of genera have focused on the rearrangements of continental masses that occurred since the appearance of seagrasses 100 million years ago. The disjunct distributions of many genera are explained by hypothesizing a much wider earlier distribution, followed by local extinctions to yield the present distribution (Larkum & Den Hartog, 1989). Indeed, the genus *Zostera*, which may include *Hetero-zostera* species as well (Les *et al.*, 1997), presents a remarkably cosmo-politan distribution, extending in latitude from 72° N to 46° S (with a gap in the equatorial region), and being present in all oceans – except the Southern Ocean (Fig. 1.1). Likewise, the distribution of other species may have been much broader in the past (Larkum & Den Hartog, 1989). The observation that the highest species richness is found in the Indo-

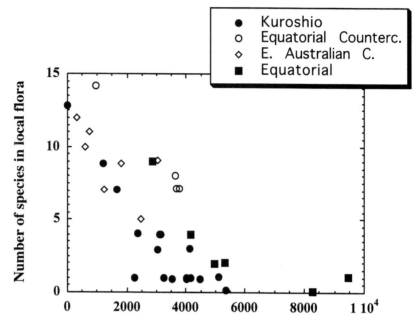

Fig. 1.2. The decline in species presence with distance along the main current systems from the Malaysian centre of species richness in the Indo-Pacific region. (From data in Mukai, 1993.)

Pacific region has also led to the hypothesis that this area is the centre of origin of seagrasses, from which they spread to other regions of the world. This hypothesis is supported by the pattern of declining species numbers with distance from this hypothetical centre along the major currents in the area (Fig. 1.2). It is evident, however, that all these hypotheses have been formulated *ad hoc*, and that they can only be rigorously tested to confirm or reject hypothetical extinctions through a substantial increase of the fossil record available and, possibly, through use of molecular phylogenetic techniques.

The examination of the biogeographic boundaries between seagrass floras provides a yet unexplored tool to try to explain the factors controlling seagrass distribution. Boundaries between seagrass floras are sometimes sharp, rather than gradual, as is the case for *Posidonia oceanica*, whose western distributional limit in the Mediterranean is closely associated with the Almería–Oran density front – a boundary between water masses of different density which separates the Atlantic from the Mediterranean surface waters (Marbà *et al.*, 1996). A thermal front separating tropical Australian from colder south-west Australian waters has also been identified to represent the boundary between the temperate and tropical Australian floras. The frontal region between the intrusion of low-salinity Chinese waters along the north coast of Vietnam and the warm waters of the South China Sea, south of the city of Hue (Vietnam) appears to represent the biogeographical boundary between the West Pacific flora and the Indo-Pacific floras. Oceanographic fronts can act as boundaries between seagrass floras by imposing sharp changes in growth conditions, such as water temperature, salinity and nutrient availability. Yet these gradients, although sharp, typically involve only a 2–3 °C change, and are, therefore, small relative to the known tolerance of seagrasses and the ranges that are experienced by individual species across their distributional area. Presumably, the role of density fronts as physical barriers that would prevent the dispersal of seagrass propagules is more important in accounting for their relation with seagrass bio-geographic boundaries. A similar effect of density fronts has been demonstrated for other marine organisms. Density fronts are themselves subject to changes in location and strength in relation to changes in oceanic circulation, as documented by paleoceanographic records. It is, therefore, possible that changes in seagrass distribution and past extinc-tions and expansions may have been associated with parallel changes in marine circulation. The ability to maintain a genetic exchange across a fragmented distribution would be dependent on the capacity of seagrass propagules or vegetative fragments for long-range transport, a capacity that is yet to be quantified.

1.5 Diversity of seagrass meadow

Most seagrass meadows are monospecific (Fig. 1.3), particularly those in the temperate zone, but this holds also for tropical and subtropical areas with multispecific floras (Figs. 1.3–1.5). Although seagrass species richness, the number of species present in any one meadow, can be significant, seagrass species diversity, i.e. the evenness of their contri-bution to the community, is typically low, as even in tropical multispecific meadows the distribution of biomass is generally skewed, with one or a few species comprising most of the biomass of the community (Terrados *et al.*, 1997). Hence, the diversity of seagrass meadows, computed as the Shannon–Weaver index (H, where $H = -\Sigma\ p_i \log_2 p_i$, where p_i is the fractional contribution of species *i* to the community), is typically 0 (i.e.

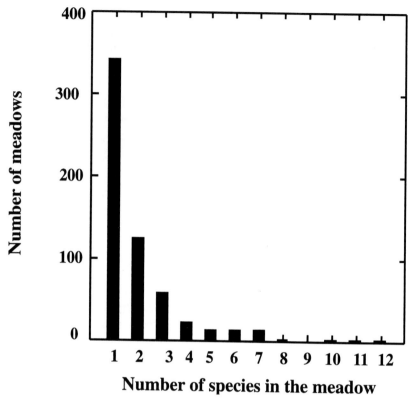

Fig. 1.3. The frequency distribution of the number of species present in 596 seagrass meadows reported in the literature. (Duarte, 2000.)

one species only), but values up to 1.56 have been reported for the Indo-Pacific region (Duarte, 2000). Hence, the species diversity of even the most diverse seagrass communities is low compared with that of terrestrial plant communities or communities of other marine organisms, which typically range between 1 and 4 (Margalef, 1980). The major part of the biological diversity of seagrass meadows, however, is contributed by the rich associated fauna and algal floras, and not by the seagrasses themselves (Chapter 6). The meadows with the richest species diversity are found in the Indo-Pacific area and the Red Sea (Fig. 1.4), where mixed meadows are abundant, containing up to 12 co-occurring species (Duarte, 2000). As a consequence, there is a tendency for a decline in the species richness and diversity of seagrass meadows from the equatorial zone to higher latitudes (Fig. 1.5). This decline is asymmetric, but includes a local maximum in species richness in the austral subtropic,

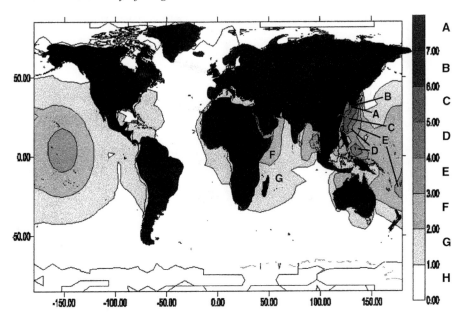

Fig. 1.4. Contour isolines of seagrass species richness generated from data on 596 seagrass meadows reported in the literature. (Duarte, 2000.)

corresponding to the high species richness present in some West Australian meadows (Kirkman & Walker, 1989). The geographical pattern of seagrass species richness is parallel to that of corals and mangroves (Heck & McCoy, 1979), which show a similar distribution of areas of high and low species richness in the tropics, suggesting that the constraints and processes responsible for the development and maintenance of species diversity of these taxa have been linked throughout their evolutionary history.

Studies aimed at describing patterns of species richness within floras are still few. These include the demonstration of a sharp reduction in species richness with increasing siltation in South-East Asian seagrass meadows (Fig. 1.6), where seagrass meadows range from monospecific (or are lost all together), in highly silted areas to containing up to seven species in the most pristine conditions. In addition, disturbance due to heavy grazing by vertebrates, such as dugongs, selects for fast-growing seagrasses, which may yield monogeneric meadows of *Halophila* (Preen, 1995), the fastest growing seagrass (cf. Duarte, 1991a). Experimental burial of seagrass meadows has also been reported to result in a reduced species diversity (Duarte *et al.*, 1997). It does appear, therefore, that

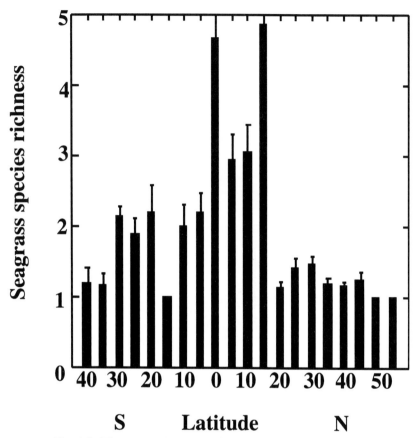

Fig. 1.5. The mean (± SE) angiosperm species richness in seagrass meadows growing at different latitudinal ranges. (Duarte, 2000.)

seagrass species richness is negatively affected by disturbance (Terrados *et al.*, 1997), since the number of species able to tolerate disturbance decreases as the magnitude of the disturbance increases (Duarte, 2000).

The extent of genetic diversity within seagrass populations is highly variable. Genetic diversity is so low in some populations that low-power techniques, such as isozyme analysis, cannot discern any significant patterns. The hypothesis that *Posidonia oceanica* meadows spreading over many kilometres could be a single clone, for instance, could not be rejected using RAPDs (randomly amplified polymorphism detection, Procaccini *et al.*, 1996). New results using more powerful techniques, such as comparison of highly variable DNA sequences in microsatellite DNA, have now provided evidence of genetic diversity hidden to less

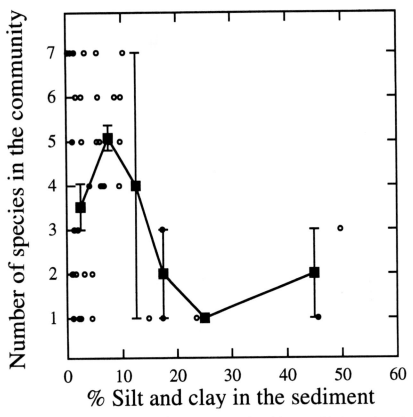

Fig. 1.6. The decline in seagrass species richness with increasing silt content in South East Asian seagrass meadows. (Terrados *et al.*, 1997.)

powerful techniques. Local studies of genetic diversity have become available for only a few species, indicating high genetic diversity in some species (e.g. *Posidonia australis*, Waycott *et al.*, 1997; *Thalassia testudinum*, Kirsten *et al.*, 1998; Schlueter & Guttman, 1998; *Halodule uninervis*, McMillan, 1982; and *Zostera marina*, Fain *et al.*, 1992; Alberte *et al.* 1994), while very low levels of genetic variability have been demonstrated for other species or populations (e.g. *Amphibolis antarctica*, cf. Waycott *et al.* 1996; and *Posidonia oceanica*, cf. Capiomont *et al.*, 1996; Procaccini *et al.*, 1996; Procaccini & Mazella 1998; *Zostera marina*, cf. Reusch *et al.*, 1999). Whether the perceived difference between seagrass populations of high- and low-genetic diversity reflects an objective difference among the plants, or whether it is a product of contrasting resolution in the various methods used to assess genetic diversity, cannot yet be resolved. The view that gene flow between populations, even those separated by only a few kilometres, is relatively low, also seems to lose power with the increasing number of studies on seagrass population genetics (Duarte, 2000). As a result, whether the level of genetic uniformity of seagrasses is greater than that typical of terrestrial plants is still unclear, and the view that the low levels of speciation in seagrasses derives from inefficient sexual reproduction is not yet fully supported. However, the genetic variation of highly clonal populations, such as *Posidonia oceanica*, is greater near their biogeographical boundary (Duarte, 2000).

1.6 The habitat requirements of seagrasses

The preceding discussion on seagrass distribution and diversity does not predict whether any particular littoral area will support seagrass cover. For instance, populations of individual seagrass species may be fragmented, with gaps between neighbour populations exceeding hundreds of kilometres due to unsuitable habitat conditions. The definition of the habitat requirements of seagrasses has been found to provide a useful tool to manage seagrass resources and prevent their decline from sensitive areas, and is, therefore, an active field of research. Being rooted phototrophs with obligate marine life cycles, the four most obvious requirements of seagrasses are a marine environment, adequate rooting substrate, sufficient immersion in seawater and illumination to maintain growth. While the light requirements of seagrasses have been investigated in some detail (see section 4.2), the sediment conditions required to support adequate seagrass growth as well as their immersion requirements are only now becoming quantitatively defined.

Most seagrass species can tolerate a wide range of salinity, from full-strength seawater to either brackish or hypersaline waters. Seagrasses are often found growing in estuaries, in contact with brackish-water plants, such as *Ruppia* spp. – which some authors include in the seagrass group – and *Potamogeton* spp. The area of contact between seagrasses and salt-tolerant freshwater macrophytes is found at around 10‰ in a number of estuaries, and seagrass growth declines at salinities in excess of about 45‰, where mortality occurs (e.g. Quammen & Onuf, 1993). Seagrass tissues suffer osmotic stress at low and high salinity, leading to loss of functionality, and they eventually become necrotic and die (Biebl & McRoy, 1971). While germination experiments have repeatedly shown seagrass seeds to germinate best at very low salinities (down to 4.5‰), experiments with *Zostera marina* have shown that the survival of

Fig. 1.7. *Phyllospadix torreyi* growing intertidally in the Baja California Peninsula. (Photograph by J. Terrados.)

seedlings is highest at salinities close to full strength seawater (Biebl & McRoy, 1971).

The depth limit of seagrasses is set by the compensation irradiance for growth, or the irradiance required to provide sufficient carbon gains to compensate for carbon losses (see section 4.2). The light requirement for seagrass growth is typically defined as the percentage of surface irradiance that needs be received by the plants to grow, which ranges between 4% and 29% (Dennison *et al.*, 1993), with an average of about 11% of the irradiance incident just below the water surface (Duarte, 1991b). These light requirements are greater than those generally observed for other marine phototrophs, such as macroalgae and microalgae (Duarte, 1995).

Whereas the study of the deep, downslope limit of seagrasses has received considerable attention, the factors determining their shallow, upper growth limit has been largely neglected. Seagrasses can develop large intertidal populations, although not all seagrass species are able to withstand exposure to air, and only a few species are common in the intertidal. *Zostera noltii* is particularly widespread in the intertidal zone of the coasts of Western Europe and North-West Africa. *Phyllospadix* species develop intertidally along rocky shores of the North Pacific (Fig. 1.7), although species-specific differences in their resistance to desiccation have led to segregation of *Phyllospadix* species in the intertidal (Ramírez-Garcia *et al.*, 1998). *Halophila* species also form large intertidal meadows on mud flats in the Indo-Pacific region. Intertidal seagrasses are best able to resist exposure to air when forming dense, continuous populations, as their leaves, lying flat on the sediment surface, retain water (e.g. 174 L m^{-2} in a *Thalassia testudinum* meadow; Powell & Schaffner, 1991), thereby avoiding desiccation. Species that typically grow subtidally may occasionally grow in the lower intertidal, but become stunted, with shoot sizes less than half the size of the adjacent, subtidal plants. The stress experienced by plants exposed to air for long periods is not limited to desiccation, but also to photodamage by high irradiance. Indeed, high UV levels cause stress, depressing photosynthesis, as has been shown experimentally (Dawson & Dennison, 1996). Intertidal seagrasses often show leaves with red spots, corresponding to high concentrations of UV-blocking pigments.

The desiccation limit of seagrasses may not be the actual shallow, upslope limit of seagrasses in areas where disturbance by wave action or ice scour is significant. The upslope limit of seagrasses is deeper in exposed, high-energy environments, compared with sheltered bays, although quantitative models describing this relationship, comparable to that between incident light and the downslope limit of seagrasses (e.g., Duarte, 1991b), are yet to be developed. The vertical and horizontal distribution of seagrasses, furthermore, is constrained by the availability of appropriate substrate for growth. In addition, growth is often restricted due to the occupation of space by other benthic communities. The limitation of seagrass communities to depths much shallower than expected from the penetration of light is particularly evident in pristine tropical waters, where seagrass meadows are confined to the lagoons enclosed by coral reefs, which completely cover the substrate at depth. Similarly, mangrove communities may extend over the intertidal zone of tropical waters, shading out seagrasses from their potential shallower upslope limit.

Where space is available, seagrass populations can only develop if the substrate is suitable. Most seagrass species are confined to sandy to muddy sediments, although some species can grow over rock. This capacity is particularly remarkable for *Phyllospadix* spp. (Fig. 1.8), which form dense stands over rocky shelfs on the Pacific coast (Den Hartog, 1970). Other species able to grow over rock include *Posidonia oceanica* and *Thalassodendron* spp. All of the species able to grow over rock are characterized by sturdy roots, which penetrate into the crevices of the

Fig. 1.8. *Phyllospadix torreyi* growing on boulders in the south California coast. (Photograph by J. Terrados.)

underlying rocks, effectively anchoring the plants. Most seagrass species do, however, grow over sandy to muddy sediments, which are easily penetrated by seagrass roots. High mobility of these fine sediments, in which currents and wave-induced bedload transport generate large sand ripples and sand waves, renders them unsuitable to support plant growth. These processes cause successive burial and erosion, which may cause seagrass mortality, depending on the size and frequency of these events relative to the life history and growth capacity of the species (cf. Chapter 3). Hence, highly mobile, but otherwise suitable, sandy sediments may be bare of seagrass cover.

Marine sediments can be hostile habitats for plant life, particularly where inputs of organic matter are excessive. High inputs of organic matter stimulate bacterial activity, raising the anoxic layer closer to the sediment surface and leading to the development of bacterial communities with metabolic pathways that result in the accumulation of phytotoxic compounds, such as sulphide (Hemminga, 1998). Seagrasses may counterbalance these stresses by pumping oxygen through their roots into the sediments, thereby maintaining a relatively oxidized rhizosphere, which has been estimated to be 80 μm thick in *Cymodocea rotundata* (Pedersen *et al.*, 1998). The organic matter concentrations of

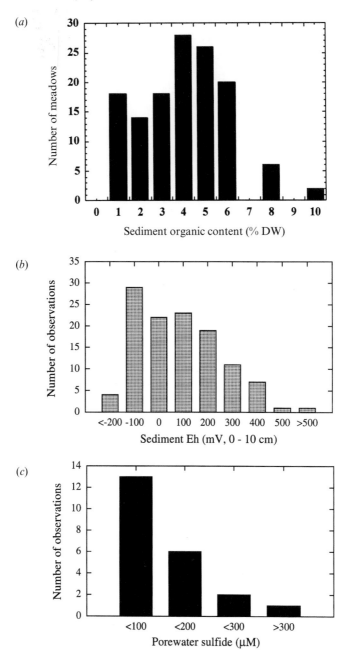

Fig. 1.9. Frequency distribution of (*a*) sediment organic matter content in 138 South-East Asian seagrass meadows (Kamp-Nielsen, unpubl. results), (*b*) redox potential, and (*c*) sulphide concentration in seagrass sediments. *b* and c reproduced from Terrados *et al.* (1997).

sediments supporting seagrass growth is generally < 6% of the dry weight (Fig. 1.9*a*). Sediments with higher organic matter concentrations are likely to support high bacterial activity, driving the sediment to a reduced status, reflected in highly negative redox potential. Seagrasses grow in sediments with redox potentials spanning from highly oxidized to moderately reduced (> − 100 mV, Fig. 1.9*b*). Seagrass growth at yet more reduced (i.e. more negative) redox conditions has been rarely observed (Terrados *et al.*, 1999). Sulphate-reducing and methane producing-bacteria develop as redox potentials fall below − 120 mV and − 140 mV, respectively (Stumm & Morgan, 1991), where they produce sulphide and methane, both of which may be toxic to higher plants. Seagrasses may survive moderate levels of such toxic compounds through different metabolic adaptations, but seagrass meadows have been reported to grow in sediments with sulphide concentrations generally less than 300 μM (Fig. 1.9*c*).

The negative influence of organic matter inputs and the associated reduced sediment conditions and accumulation of phytotoxins on seagrasses have been tested experimentally. Similar inputs of organic matter and increases in sulphide concentrations had remarkably different effects on the growth of different seagrass species (Terrados *et al.*, 1999). These results suggest that the negative effects of reduced sediments are far from simple, and that our present understanding of the growth of seagrasses in anoxic sediments is insufficient to predict tolerance levels reliably.

The negative effects of excessive inputs of organic matter on seagrass growth have led to the notion that highly productive seagrass meadows may 'poison' themselves by driving the sediment conditions to stressful levels (see section 7.5). Indeed, the presence of seagrass meadows does increase organic inputs to the sediments, not only through their own detritus but also through the trapping of suspended particles (Ward *et al.*, 1984; Duarte *et al.*, 1999; Gacia *et al.*, 1999). There are other stress factors that may limit the habitat of seagrasses, such as high nitrate and ammonium concentrations in the water column, which have been reported to have negative effects on seagrasses (Burkholder *et al.*, 1992; Van Katwijk *et al.*, 1997). Seagrasses appear to be relatively resistant to contaminants (e.g. Kenworthy *et al.*, 1993), but they are unlikely to inhabit highly polluted habitats, which would probably have a rather reduced transparency as well (see section 7.5).

1.7 Seagrass abundance and productivity in the global context

The global biomass and production of benthic macrophytes, seagrasses and macroalgae are more difficult to estimate than those of phytoplankton and mangroves, of which the area and biomass can be delineated from satellite images. Local estimates have been derived in shallow areas, where seagrass cover can also be assessed from satellite imagery, or from large-scale applications of side-scan sonar. For instance, the area covered by seagrasses in Florida Bay has been estimated to be 14 662 km^2 (Zieman *et al.*, 1989), while the area covered by *Posidonia oceanica* on the Spanish Mediterranean coast has been estimated, using a combination of approaches, as 2809 km^2 (calculated from information in Mas *et al.*, 1993). However, even though seagrass cover has been estimated over large areas, this exercise has been done in only a few regions, so that it is impossible to estimate the global area covered by seagrasses by summation of local estimates. Thus, present estimates of the area covered by marine benthic macrophytes are based on indirect calculations. These estimates assign a total global area of about 0.6×10^6 km^2 to seagrasses (Charpy-Roubaud & Sournia, 1990), which is equivalent to 10% of the coastal ocean (or 0.15% of the global ocean). The area covered by seagrasses is, thus, comparable to that of coral reefs, macroalgae, and mangroves, although the precision of this estimate is unknown.

Seagrass meadows often support high biomass, with an average of about 460 g DW m^{-2}, an estimate derived from a compilation of biomass estimates of seagrass meadows throughout the world (Duarte & Chiscano, 1999). In particular, dense populations of the genera *Amphibolis*, *Phyllospadix* and *Posidonia* tend to support high biomasses, while those of *Halophila* tend to support a much lower biomass (Fig. 1.10). Because of their high biomass compared with that of phytoplankton, the global biomass of seagrasses is disproportionately large relative to the area they cover (Smith, 1981), and represents about 1% of the total biomass of marine plants.

Seagrass meadows are also productive ecosystems. The estimates of primary production of seagrass meadows indicate an average net production of about 1012 g DW m^{-2} yr^{-1}, when the production of roots is considered (Duarte & Chiscano, 1999). This estimate of the average net primary production of seagrass meadows places them amongst the most productive ecosystems in the biosphere (Table 1.2). The most productive seagrass populations appear to be those of *Phyllospadix* spp. along the Pacific coasts, and those of *Posidonia oceanica* (Fig. 1.10), as well as the

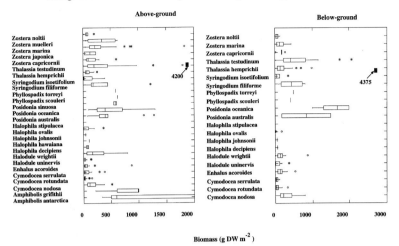

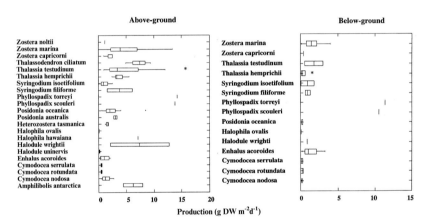

Fig. 1.10. The distribution of maximum seagrass biomass (*top*) and production (*bottom*) of different seagrass species. Boxes encompass 25% and 75% quartiles, the central line represents the median, and bars encompass 95% of the values. Asterisks and open circles indicate observations outside the 95% limits. (Duarte & Chiscano, 1999.)

dense mixed meadows of the Indo-Pacific region (e.g. Vermaat *et al.*, 1995). These estimates represent, when scaled to the estimated global cover of seagrasses, a contribution to marine primary production of 0.6×10^{15} g C yr^{-1}, or about 1.13% of the total marine primary production (Duarte & Cebrián, 1996). While most of the production of phytoplankton is used up in the marine ecosystem, that of seagrasses is not, and most of their primary production is either stored in the sediments or exported to neighbouring ecosystems (Duarte & Cebrián,

Table 1.2. *The average primary production of seagrass populations compared with that of other ecosystems*

Community	Biomass (g DW m^{-2})	Production (g DW m^{-2} d^{-1})	Reference
Forests			
Tropical	45000	5.2	Whittaker (1975)
Temperate	35000	3.4	Whittaker (1975)
Boreal	20000	2.2	Whittaker (1975)
Grasslands			
Savanna	4000	2.4	Whittaker (1975)
Temperate	1600	1.6	Whittaker (1975)
Tundra and alpine			Whittaker (1975)
Swamp and marshes	15000	5.5	Whittaker (1975)
Cultivated land	1000	1.8	Whittaker (1975)
Phytoplankton		0.35	Cebrián & Duarte (1994)
Microphytobenthos	9.2	0.13	Charpy-Robaud & Sournia (1990)
Coral reefs	2000	0.8	B: Whittaker (1975)
			P: Crossland *et al.* (1991)
Macroalgae	40.7	1.0	B: Cebrián & Duarte (1994)
			P: Charpy-Robaud & Sournia (1990)
Marsh plants	767	3.0	B: Cebrián & Duarte (1994)
			P: Woodwell *et al.* (1973)
Mangroves		2.7	P: Lugo *et al.* (1988)
Seagrasses	461	2.7	Duarte & Chiscano (1999)

B, biomass; P, production
Source: Adapted from Duarte & Chiscano (1999).

1996; Duarte & Agustí, 1998). Current knowledge indicates that about 16% of seagrass production is stored in the sediments, which – if entering long-term burial – represents a net sink of carbon in the ecosystem. Hence, the carbon stored in the sediments annually by seagrasses is estimated to be in the order of 0.16×10^{15} g C yr^{-1}, or about 15% of the total carbon storage in marine ecosystems (Duarte & Cebrián, 1996; Duarte & Chiscano, 1999). Hence, seagrasses are important components of the marine carbon cycle, being responsible for a significant fraction of the net CO_2 uptake by marine biota.

1.8 Concluding remarks

Despite the reduced membership of the seagrass flora, there are important uncertainties as to the taxonomic status and the number of extant seagrass species. The limited number of seagrasses species compared with the richness of the angiosperm flora present in freshwater and terrestrial environments suggests important constraints on the colonization of the marine environments by angiosperms. These constraints are reflected in the well-defined habitat that seagrasses occupy, which in turn reflects the conditions necessary to support angiosperm life and functions in the sea. The seagrass habitat, however, is, on a global scale, very extensive, allowing seagrasses to play an important role in maintaining biological productivity and biogeochemical cycles in the sea.

1.9 References

Ackerman, J.D. (1998). Is the limited diversity of higher plants in marine systems the result of biophysical limitations for reproduction or evolutionary constraints? *Functional Ecology*, **12**, 975.

Alberte, R.S., Suba, G.K., Procaccini, G., Zimmerman, R.C. & Fain, S.R. (1994). Assessment of genetic diverstiy of seagrass populations using DNA fingerprinting: implications for population stability and management. *Proceedings of the National Academy of Sciences USA*, **91**, 1049–53.

Arber, A. (1920). *Water plants: a study of aquatic angiosperms*. Cambridge: Cambridge University Press. Reprinted 1972 by Verlag von J. Cramer, Lehre.

Biebl, R. & McRoy, C.P. (1971). Plasmatic resistance and rate of respiration and photosynthesis of *Zostera marina* at different salinities and temperatures. *Marine Biology*, **8**, 48–56.

Burkholder, J.M., Mason, K.M. & Glasow Jr., H.B. (1992). Water-column nitrate enrichment promotes decline of eelgrass *Zostera marina*: evidence from seasonal mesocosm experiments. *Marine Ecology Progress Series*, **81**, 163–78.

Capiomont, A., Sandmeier, M., Cayé, G. & Meinesz, A. (1996). Enzyme polymorphism in *Posidonia oceanica*, a seagrass endemic to the Mediterranean. *Aquatic Botany*, **54**, 265–77.

Cebrián, J. & Duarte, C.M. (1994). The dependence of herbivory on growth rate in natural plant communities. *Functional Ecology*, **8**, 518–25.

Charpy-Roubaud, C. & Sournia, A., (1990). The comparative estimation of phytoplanktonic and microphytobenthic production in the oceans. *Marine Microbial Food Webs*, **4**, 31–57.

Crossland, C.J., Hatcher, B.G. & Smith, S.V. (1991). Role of coral reefs in global ocean production. *Coral Reefs*, **10**, 55–64.

Dawson, S.P. & Dennison, W.C. (1996). Effects of ultraviolet and photosynthetically active radiation on five seagrass species. *Marine Biology*, **125**, 629–38.

Den Hartog, C. (1970). *The Seagrasses of the World*. Amsterdam: North Holland.

Dennison, W.C., Orth, R.J., Moore, K.A., Stevenson, J.C., Carter, V., Kollar, S., Bergstrom, P.W. & Batiuk, R.A. (1993). Assessing water quality with submersed aquatic vegetation. *BioScience*, **43**, 86–94.

Drew, E.A. (1983). Sugars, cyclitols and seagrass phylogeny. *Aquatic Botany*, **15**, 387–408.

Duarte, C.M. (1991a). Allometric scaling of seagrass form and productivity. *Marine Ecology Progress Series*, **77**, 289–300.

Duarte, C.M. (1991b). Seagrass depth limits. *Aquatic Botany*, **40**, 363–77.

Duarte, C.M. (1995). Submerged aquatic vegetation in relation to different nutrient regimes. *Ophelia*, **41**, 87–112.

Duarte, C.M. (2000). Benthic ecosystems: seagrasses. In *Encyclopedia of Biodiversity* ed. S.L. Levin. San Diego: Academic Press (in press).

Duarte, C.M. & Agustí, S. (1998). The CO_2 balance of unproductive aquatic ecosystems. *Science*, **281**, 234–6.

Duarte, C.M. & Cebrián, J. (1996). The fate of marine autotrophic production. *Limnology and Oceanography*, **41**, 1758–66.

Duarte, C.M. & Chiscano, C.L. (1999). Seagrass biomass and production: a reassessment. *Aquatic Botany*, **65**, 159–74.

Duarte, C.M., Terrados, J., Agawin, N.S.W., Fortes, M.D., Bach, S. & Kenworthy, W.J. (1997). Response of a mixed Philippine seagrass meadow to experimental burial. *Marine Ecology Progress Series*, **147**, 285–94.

Duarte, C.M., Benavent, E. & Sánchez, M.C. (1999). The microcosm of particles within seagrass (*Posidonia oceanica*) canopies. *Marine Ecology Progress Series*, **181**, 289–95.

Fain, S.R., De Tomaso, A. & Alberte, R.S. (1992). Characterization of disjunct populations of *Zostera marina* (eelgrass) from California: genetic differences resolved by restriction-fragment length polymorphisms. *Marine Biology*, **112**, 683–9.

Gacia, E., Duarte, C.M. & Granata, T. (1999). An approach to the measurement of particle flux and sediment retention within seagrass (*Posidonia oceanica*) meadows. *Aquatic Botany* **65**, 255–68.

Gagnon, P.S., Vadas, R.L., Burdick, D.B. & May, B. (1980). Genetic identity of annual and perennial forms of *Zostera marina* L. *Aquatic Botany*, **8**, 157–62.

Heck Jr., K.L. & McCoy, E.D. (1979). Biogeography of seagrasses: evidence from associated organisms. *New Zealand Department of Scientific and Industrial Research Information Series*, **137**, 109–28.

Hemminga, M.A. (1998). The root/rhizome system of seagrasses: an asset and a burden. *Journal of Sea Research*, **39**, 183–96.

Kenworthy, W.J., Durako, M.J., Fatemy, S.M.R., Valavi, H. & Thayer, G.W.

(1993). Ecology of seagrasses in North-eastern Saudi Arabia one year after the Gulf War oil spill. *Marine Pollution Bulletin*, **27**, 213–22.

Kirkman, H. & Walker, D.I. (1989). Regional studies – Western Australian seagrass. In *Biology of Seagrasses*, ed. A.W.D. Larkum, A.J. McComb & S.A. Shepherd, pp. 157–81. New York: Elsevier.

Kirsten, J.H., Dawes, C.J. & Cochrane, B.J. (1998). Ramdomly amplified polymorphism detection (RAPD) reveals high genetic diversity in *Thalassia testudinum* Banks ex König (Turtlegrass). *Aquatic Botany*, **61**, 269–87.

Larkum, A.W.D. & den Hartog, C. (1989). Evolution and biogeography of seagrasses. In *Biology of Seagrasses*, ed. A.W.D. Larkum, A.J. McComb & S.A. Shepherd, pp. 112–56. Elsevier, New York.

Les, D.H., Cleland, M.A. & Waycott, M.A. (1997). Phylogenetic studies in Alismatidae. II. Evolution of marine angiosperms (seagrasses) and hydrophily. *Systematic Botany*, **22**, 443–63.

Lugo, A.E., Brown, S. & Brinson, M.M. (1988). Forested wetlands in freshwater and salt-water environments. *Limnology and Oceanography*, **33**, 894–909.

Marbà, N., Duarte, C.M., Cebrián, J., Enríquez, E., Gallegos, M.E., Olesen, B. & Sand-Jensen, K. (1996). Growth and population dynamics of *Posidonia oceanica* on the Spanish Mediterranean coast: elucidating seagrass decline. *Marine Ecology Progress Series*, **137**, 203–13.

Margalef, R. (1980). *Ecología*. Barcelona: Omega.

Mas J., Franco, I. & Barcala, E. (1993). Primera aproximación a la cartografía de las praderas de *Posidonia oceanica* en las costas mediterráneas espanolas. Factores de alteración y de regresión. Legislación. *Publicaciones Especiales del Instituto Español de Oceanografía*, **11**, 111–22.

McMillan, C. (1982). Isozymes in seagrasses. *Aquatic Botany*, **14**, 231–43.

Mukai, H. (1993). Biogeography of the tropical seagrasses in the Western Pacific. *Australian Journal of Marine Freshwater Research*, **44**, 1– 17.

Pedersen, O., Borum, J., Duarte, C.M. & Fortes, M.D. (1998). Oxygen dynamics in the rhizosphere of *Cymodocea rotundata*. *Marine Ecology Progress Series*, **169**, 283–8.

Phillips, R.C. & Meñez, E.G. (1988). *Seagrasses*. Smithsonian Contrib. Mar. Sci. 34. Washington DC: Smithsonian Institution.

Powell, G.V.N. & Schaffner, F.C. (1991). Water trapping by seagrasses occupying bank habitats in Florida Bay. *Estuarine Coastal and Shelf Science*, **32**, 43–60.

Preen, A. (1995). Impacts of dugong foraging on seagrass habitats: observational and experimental evidence for cultivation grazing. *Marine Ecology Progress Series*, **124**, 201–13.

Procaccini, G. & Mazella, L. (1998). Population genetic structure and gene flow in the seagrass *Posidonia oceanica* assessed using microsatellite analysis. *Marine Ecology Progress Series*, **168**, 133–41.

Procaccini, G., Alberte, R.S. & Mazella, L. (1996). Genetic structure of the seagrass *Posidonia oceanica* in the Western Mediterranean: ecological implications. *Marine Ecology Progress Series*, **140**, 153–60.

Quammen, M.L. & Onuf, C.P. (1993). Laguna Madre: Seagrass changes continue decades after salinity reduction. *Estuaries*, **16**, 302–10.

Ramírez-García, P., Lot, A., Duarte, C.M., Terrados, J. & Agawin, N.S.R. (1998). Bathymetric distribution, biomass and growth dynamics of intertidal *Phyllospadix scouleri* and *Phyllospadix torreyi* in Baja California (Mexico). *Marine Ecology Progress Series*, **173**, 13–23.

Raven, J.A. (1977). The evolution of vascular land plants in relation to supracellular transport processes. *Advances in Botanical Research*, **5**, 153–219.

Reusch, T.B.H., Borström, C., Stam, W.T. & Olsen, J.L. (1999). An ancient eelgrass clone in the Baltic. *Marine Ecology Progress Series*, **183**, 301–4.

Robertson, J.I. & Mann, K.H. (1984). Disturbance by ice and life-history adaptations of the seagrass *Zostera marina*. *Marine Biology*, **80**, 131–41.

Schlueter, M.A. & Guttman, S.I. (1998). Gene flow and genetic diversity of turtle grass, *Thalassia testudinum* Banks ex König, in the lower Florida Keys. *Aquatic Botany*, **61**, 147–64.

Smith, S.V. (1981). Marine macrophytes as a global carbon sink. *Science*, **211**, 838–40.

Stumm, W.S. & Morgan, J.J. (1991). *Aquatic Chemistry*. New York: John Wiley.

Terrados, J., Duarte, C.M., Fortes, M.D., Borum, J., Agawin, N.S.R., Bach, S., Thampanya, U., Kamp-Nielsen, L., Kenworthy, W.J., Geertz-Hansen, O. & Vermaat, J. (1997). Changes in community structure and biomass of seagrass communities along gradients of siltation in SE Asia. *Estuarine, Coastal and Shelf Science*, **46**, 757–68.

Terrados, J., Duarte, C.M., Kamp-Nielsen, L., Agawin, N.S.R., Gacia, E., Lacap, D., Fortes, M.D., Borum, J., Lubanski, M. & Greve, T. (1999). Are seagrass growth and survival affected by reducing conditions in the sediment? *Aquatic Botany*, **65**, 175–98.

Van der Hage, J.C.H. (1996). Why are there no insects and so few higher plants in the sea? New thoughts on an old problem. *Functional Ecology*, **10**, 546–7.

Van Katwijk, M.M., Vergeer, L.H.T., Schmitz, G.H.W. & Roelofs, J.G.M. (1997). Ammonium toxicity in eelgrass *Zostera marina*. *Marine Ecology Progress Series*, **157**, 159–73.

Vermaat, J.E., Fortes, M.D., Agawin, N.S.R., Duarte, C.M., Marbà, N. & Uri, J. (1995). Meadow maintenance, growth, and productivity of a mixed Philippine seagrass bed. *Marine Ecology Progress Series*, **124**, 215–55.

Ward, L.G., Kemp, W.M. & Boynton, W.R. (1984). The influence of waves and seagrass communities on suspended particulates in an estuarine embayment. *Marine Geology*, **59**, 85–103.

Waycott, M. (1999) Mating systems and population genetics of marine angiosperms (seagrasses). In *Systematics and Evolution of Monocots*, Vol. 1 of Proceedings. 2nd International Conference on Comparative Biology of Monocotyledons, ed. K.L. Wilson & D. Morrison. Sydney: CSIRO Publishing.

Waycott, M., Walker, D.I. & James, S.H. (1996). Genetic uniformity in a dioecious seagrass, *Amphibolis antarctica*. *Heredity*, **76**, 578–85.

Waycott, M., James, S.H. & Walker, D.I. (1997). Genetic variation within and between populations of *Posidonia australis*, a hydrophilous, clonal seagrass. *Heredity*, **79**, 408–12.

Whittaker, R.H. (1975). *Communities and Ecosystems*, 2nd edn. London: Macmillan.

Woodwell, G.M., Rich, P.H. & Hall, C.S.A. (1973). Carbon in estuaries. In *Carbon in the Biosphere*, ed. G.M. Woodwell & E.V. Pecan, pp. 221–40. U.S. AEC.

Zieman, J.C., Fourqurean, J.W. & Iverson, R.L. (1989). Distribution, abundance, and productivity of seagrasses and macroalgae in Florida Bay. *Bulletin of Marine Science*, **44**, 292–311.

2

Seagrass architectural features

2.1 Introduction

The few members of the angiosperm flora that have succeeded in adapting to submersed life in the sea share a common architecture, all species being clonal, rhizomatous plants. This clonal nature has been interpreted as a necessary adaptation for angiosperm growth in the high-energy marine environment (Sculthorpe, 1967). A consequence of the clonal nature of the seagrasses is that they display a highly ordered growth programme (Tomlinson, 1974), developed through the regular addition of the basic set of modules. Thus, a general understanding of the design of seagrasses will provide insight into their growth patterns (Patriquin, 1973, 1975; Tomlinson, 1974; Sand-Jensen, 1975; Duarte & Sand-Jensen, 1990; Duarte *et al.*, 1994). Although the repertoire of architecture and associated growth programmes that seagrasses display is certainly narrow, they contain sufficient plasticity to yield order-of-magnitude variability in the clonal growth between individual shoots of seagrass species (Marbà & Duarte, 1998), as well as in their reproduction and dispersal. Even within a species, the plasticity of its growth programme and architecture allows the plants to cope with stress and heterogeneity in the environment, as has been extensively documented for land plants. This plasticity is also a central trait in the ecology of seagrasses.

In this chapter we provide a description of the basic architecture of seagrasses, the growth patterns resulting from their design, and their plasticity in growth characteristics to cope with stress and resource heterogeneity. Differences in the design of seagrasses have been found to be most useful in predicting their capacity to resist disturbance and to recover from it (Duarte *et al.*, 1994; Vermaat *et al.*, 1997), thereby helping to explain their role in the community (Duarte, 1991). Hence, the goal of this chapter is to provide the reader with the necessary details of the design and clonal growth patterns of seagrasses to understand how these are linked to their performance in nature. We shall not, however, present a thorough discussion of some aspects, such as an in-depth description of the anatomy of seagrass organs (drawings for the different species can be found in Den Hartog, 1970), which albeit important, have not produced much insight into the ecology of the species. A thorough description of the anatomy and ultrastructure of seagrass modules can be found in Kuo & McComb (1989).

2.2 Seagrass modules: rhizomes, leaves, roots, flowers, fruits

Seagrasses are modular plants composed of units (ramets), which are repeated during clonal growth. Each ramet is composed of a set of modules: a piece of rhizome, which can be either horizontal or vertical; a bundle of leaves attached to the rhizome; and a root system (Fig. 2.1). In addition, the ramets may hold flowers or fruits, depending on the timing of observation. Differences across the seagrass flora result from small variations in this basic repertoire, such as variations in the number of these modules per ramet, their size and form. The morphology of seagrasses does not present any peculiar deviations relative to those of other terrestrial monocotyledons.

2.2.1 Rhizomes

Seagrasses are rhizomatous plants, meaning that they have stems extending horizontally below the sediment surface (Figs. 2.1, 2.2). The rhizome is responsible for the extension of the clone in space, as well as for connecting neighbouring ramets, thereby maintaining integration within the clone (see below). Seagrass rhizomes are leptomorph (i.e. long and relatively thin) but contain large lacunae in the outer cortical tissue (Kuo & McComb, 1989) which allow efficient transfer of gases. They also contain vascular systems involved in the transfer of resources and hormones (Terrados-Muñoz, 1995) to maintain both physical and physiological contact. The rhizomes of small seagrass species are flexible, whereas those of large seagrass species are often strongly lignified, being almost woody in some species such as *Enhalus acoroides* and *Posidonia oceanica* (Den Hartog, 1970). The extent of lignification of the rhizomes appears to be associated with their longevity, rather than with size (cf. Klap *et al.*, 2000).

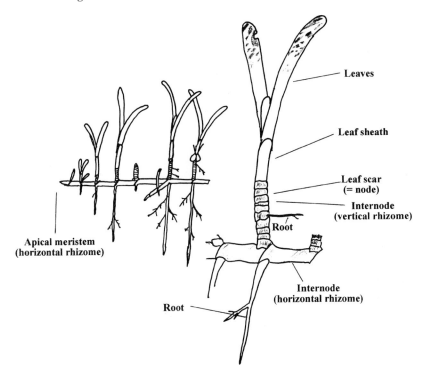

Fig. 2.1. A schematic depiction of the basic components of seagrass architecture.

Seagrass rhizomes are composed of internodes (Figs. 2.1, 2.2), the rhizome fragments between two nodes, which are the insertion points of leaves. The distinct lines identifying the nodes after leaf abscission are also referred to as 'leaf scars' (Figs. 2.1, 2.2). Rhizome internodes range widely in size, spanning about 40-fold in length and 20-fold in thickness (Table 2.1). Most of the variability in dimensions of rhizome internodes is found between species, but within-species variability in internodal length can be substantial (Table 2.1).

Many seagrass species have dimorphic rhizomes, consisting of two differentiated types: horizontal (plagiotropic) rhizomes, which typically bear relatively long internodes, and vertical (orthotropic) rhizomes, typically bearing shorter internodes (Figs. 2.1, 2.2, Table 2.1). There is some degree of confusion regarding the terms used to refer to the horizontal and vertical rhizomes. Vertical rhizomes are also referred to as short shoots in the literature. Scientists working on land plants of similar architecture would refer to them as stems, because the term shoot refers

Fig. 2.2(*a*)

(b)

(c)

Fig. 2.2. Photographs illustrating the basic modules in the architecture of seagrasses: (a) *Enhalus acoroides* (photograph by C.M. Duarte); (b) *Amphibolis antarctica* (photograph by D. Walker); (c) *Halophila decipiens* (photograph by Japar Sidik Bujang).

Table 2.1. *Mean rhizome diameter (RD, in mm), horizontal internodal length (HINTL, in mm), vertical internodal length (VINTL, in mm), horizontal rhizome length between consecutive shoots (RL, in cm), horizontal rhizome elongation rate (HE, in cm yr⁻¹), and vertical rhizome elongation rate (VE, in cm yr⁻¹) of seagrass species*. Values within brackets show the ranges of the variables per species when more than one stand has been examined

Species	RD	HINTL	VINTL	RL	HE	VE
Amphibolis antarctica	2.80	7	9 (5–12)	3.86	20 (5–35)	24 (7–32)
Amphibolis griffithii	2.19	10	10	6.16	4	16 (5–34)
Cymodocea nodosa	2.37	25 (6–53)	1.4 (0.1–2.5)	2.77 (1.1–5.5)	40 (7–204)	1.4 (0.1–16)
Cymodocea rotundata	2.44	29	2 (0.5–5)	4.8	210	1.5 (1.1–3.3)
Cymodocea serrulata	2.78	38 (35–39)	5 (4–7)	5.30	153 (34–411)	13
Enhalus acoroides[a]	14.1	5		6.68	3 (0.9–5)	
Halodule uninervis	1.37	21 (9–33)	5 (1–7)	2.7	101 (28–138)	4
Halodule wrightii	1.64	20.2	1.2	2.02	223 (18–365)	3
Halophila decipiens[a]	0.8				215	
Halophila hawaiiana[a]	1.2	10			89	
Halophila ovalis[a]	1.3	17		1.7	356 (141–574)	
Halophila stipulacea[a]	1.12	16				
Heterozostera tasmanica	1.74	20	7	2.07	103 (56–150)	9

Species					
Posidonia angustifolia	5.99	18	10	7.5	12	5 (1–11)
Posidonia australis	7.21	15 (14–17)	1.5	6	9 (9–10)	1 (0.4–6)
Posidonia oceanica	9.7	3 (1–4)	1 (0.4–2)	2.81 (1.2–5.3)	2 (1–6)	1 (0.1–4)
Posidonia sinuosa	5.5	11 (8–14)	3 (2–4)	4.47 (3–5.9)	4 (2–6)	13 (0.2–8)
Pyllospadix torreyi[a]	5				26	
Syringodium filiforme	2.77	23 (16–31)	6	3.06	123 (52–182)	4
Syringodium isoetifolium	1.74	27 (25–28)	11 (8–13)	3.7	109 (75–135)	9
Thalassodendron ciliatum	5.6	8 (6.7–10)	3	3.32 (2.7–4)	16	10 (5–10)
Thalassia hemprichii	3.63	4 (4–7)	1 (0.5–3)	6.9	54 (21–88)	3 (2–5)
Thalassodendron pachyrhizum	5.64	5	6 (5.2–7.35)	1.9	3 (3–6)	5 (38–12)
Thalassia testudinum	5.96	15 (5–43)	2 (7–18)	7 (5.8–7.1)	69 (22–152)	4 (2–20)
Zostera marina[a]	3.5	11 (9–12)		6.08	26 (22–31)	
Zostera noltii[a]	1.6	12 (3–20)		2.07	68 (10–127)	

[a] Species without differentiated horizontal and vertical rhizomes.
Source: Compiled from data in Duarte (1991) and Marbà & Duarte (1998).

to leaves plus stems (Bell, 1991). To avoid ambiguity, hereafter we shall use the term shoots to indicate leaves plus rhizome, and then discriminate between horizontal and vertical rhizomes.

Horizontal rhizomes can revert into vertical rhizomes, which leads to the cessation of horizontal growth. In turn, vertical rhizomes can branch to produce horizontal rhizomes when the apical meristem of the original horizontal rhizome dies, thereby resuming the capacity for horizontal growth. The vertical rhizomes often extend from their insertion into the horizontal rhizome up to the sediment surface (e.g. *Thalassia, Cymodocea*, Marbà *et al.*, 1994a,b; Marbà & Duarte, 1994), while those of some species (e.g. *Cymodocea serrulata, Thalassodendron, Amphibolis, Halodule* and *Syringodium* species) can extend beyond the sediment surface into the water column. While the vertical rhizomes lack flexibility in species where these develop within the sediments, those of species whose stems extend into the water column are highly flexible, allowing the wave energy to be accommodated. Vertical rhizomes are often narrow at the insertion point into the horizontal rhizomes, and their diameter increases somewhat upwards. Once their leaf-producing meristem dies, they remain attached to the rhizome but they can break at this narrow basal part, leaving a distinct scar on the horizontal rhizome.

Species with monomorphic rhizomes (i.e. with only one type of rhizome), such as *Zostera* species, produce, therefore, a leaf at each node. Species with highly differentiated dimorphic rhizomes, such as *Thalassia* and *Thalassodendron* species, only bear green leaves on the vertical rhizomes, and have scale leaves inserted on the nodes of the horizontal rhizomes instead. These scale leaves, or cataphylls, are small, chlorophyll-free remnants of leaves common in the underground stems of rhizomatous plants (Bell, 1991). Other species with dimorphic rhizomes, such as *Cymodocea* species, produce leaves on both the horizontal and vertical rhizomes. The leaves produced on the horizontal rhizomes are shorter and are rapidly shed, such that the excavated horizontal rhizomes appear either devoid of leaves or are covered by the remains of the leaf sheaths (e.g. *Posidonia*) or their fibres (e.g. *Enhalus acoroides*). These remains often confer a hairy aspect to the horizontal rhizomes (Fig. 2.2*a*). Remnants of the leaf sheaths may also be found attached to the vertical rhizomes of some species (e.g. *Thalassia* and *Posidonia*), where they remain attached for many years after the leaf blades are shed (several decades in *Posidonia oceanica*; Pergent & Pergent-Martini, 1990).

2.2.2 Leaves

Most seagrass species have strap-like leaves, which are long and relatively narrow, as most monocot leaves are. However, the leaves of some *Halophila* species, which have rounded leaves (Fig. 2.2*c*), and those of *Syringodium*, which are terete (cylindrical), depart from the general pattern. Seagrass leaves generally present parallel venation, joined by fine cross-sections. Seagrass leaves range widely in size, from the tiny leaves (as small as 1 cm in length) of some *Halophila* species, to the very long (> 1 m) and wide leaves reported for *Zostera asiatica* and *Enhalus acoroides*. Seagrass leaves also range greatly in thickness and anatomy, from bearing only two cell layers in the thin (91 μm; Enríquez *et al.*, 1992) leaves of *Halophila* species, to the multiple layers present in the thick leaves of *Enhalus* (500 μm; Enríquez *et al.*,1992). These differences are of great functional significance, for they influence the mechanical properties of the leaves (cf. Niklas, 1992, 1994), and also their efficiency in harvesting light (Enríquez *et al.*, 1992; Agustí *et al.*, 1994). Examination of the cross-section of seagrass leaves reveals a large number of areal spaces, or lacunae, surrounded by mesophyll cells. Air lacunae are continuous within all organs, with septa between modules, which allow for gas continuity but prevent flooding (Kuo & McComb, 1989). Lacunae allow the accumulation and pressure-driven flow of gases. Oxygen is transported via the lacunae from source to sink organs (Pedersen *et al.*, 1998), whereas they also accumulate respiratory carbon dioxide for internal recycling in photosynthesis (Roberts & Moriarty, 1987). These areal spaces reportedly comprise from 4% to 30% of the cross-sectional area of seagrass leaves in five Australian species (Grice *et al.*, 1996). These areal spaces confer buoyancy to the leaves, allowing them to remain upright while being flexible enough to accommodate to the currents. The xylem system is relatively reduced in seagrasses, suggesting that it supports little transport.

Seagrass leaves lack stomata, and their surfaces are covered by a thin cuticle, across which the transfer of gases and solutes must take place. In order to facilitate this exchange, the cuticle is porous or perforated (e.g. in *Thalassia testudinum*, *Posidonia* spp.; Kuo & McComb, 1989). The epidermis is rich in chloroplasts, being the main site where photosynthesis takes place, and often contains tannin cells. Leaf blades often contain large fibre cells, particularly near the leaf margins, developing strips along the longitudinal vascular bundles. These fibre strips provide the longitudinal strength for the maintenance of structural integrity under wave fields, and determine the motion of the leaves.

Seagrass leaves are inserted, through a tube-like structure in the basal part of their leaves, known as the leaf sheath, around most of the circumference of the rhizome (Fig. 2.1), similar to many monocots (Bell, 1991). The leaf sheath may fully encircle the rhizome (e.g. *Cymodocea, Zostera*), resulting in nodes that completely encircle the rhizome, or do so only partially, resulting in open nodes (e.g. *Posidonia*). The leaf sheath often lacks chlorophyll in the epidermis, but it encloses and protects the meristem, and also protects newly formed leaves, which emerge enclosed within the leaf sheath of the previously formed leaf. To better serve this protective goal, the sheath of some species often contains lignified fibre bundles in the epidermis (e.g. *Posidonia*). In such cases, the abscission of leaves, the process by which old, non-functional leaves are shed, proceeds through the junction between the leaf sheath and the blade, and the sheath remains attached to the rhizomes for some time. The greater lignin content of such leaf sheaths also implies that they are more resistant to decomposition (Enríquez *et al.*, 1993), and they, therefore, can persist for a long time attached to the rhizome, which confers the rhizomes of *Posidonia* and *Thalassia* their characteristic hairy appearance (Figs. 2.1, 2.2). The species of *Halophila* are exceptions, for their shoots carry pairs of petiolated leaves, ranging in form from oval to elongated, and lack sheaths (Fig. 2.2). The sheath fibres of *Posidonia oceanica* are rolled, after detachment, by the waves within sand ripples, forming the fibre balls that often pile on Mediterranean beaches (Fig. 2.3), and which have puzzled naturalists for centuries, giving rise to the most peculiar hypotheses as to their origin.

Leaves are often present in bundles on seagrass rhizomes. These bundles arise whenever the internodes between leaf-bearing nodes have not reached their full length, so that the nodes, and therefore the points of insertion of the leaves, are located very close to one another, resulting in the development of apparent leaf bundles. The number of leaves per bundle ranges from slightly more than one leaf in *Syringodium* to over 10 leaves in *Amphibolis*.

This is the case in both vertical and horizontal rhizomes, except for most *Halophila* species, which produce shoots with a single leaf pair, each supported by a petiole (Kuo & McComb, 1989).

2.2.3 Roots

The study of the form and growth of seagrass roots has received relatively little attention in seagrass ecology, despite their important

Fig. 2.3. Balls composed of leaf fibres of *Posidonia oceanica* are common in Mediterranean beaches. (Photograph by C.M. Duarte.)

functions and their important contribution to seagrass biomass and production (Duarte *et al.*, 1998; Duarte & Chiscano, 1999, cf. Chapter 1). Seagrass roots are, as for all monocotyledons, adventitious, since they arise from root primordia formed in the meristematic region of the rhizome apex (Figs. 2.1, 2.2), except for the primordial root formed after seed germination. Because the primordial root cannot increase in diameter as the clone grows, it soon becomes insufficient and adventitious roots must develop to provide the necessary anchoring and nutrient acquisition. Seagrass roots vary greatly in size and form, from the very thin roots of the small species (e.g. 0.18 mm in *Halodule uninervis*) to the thick roots of the large species (e.g. 3.5 mm in *Enhalus acoroides* and 1.8 mm in *Posidonia*; Duarte *et al.*, 1998). Root length also varies greatly, from the tiny roots of *Halophila* to root lengths of up to 5 m reported for *Thalassia testudinum*. Adventitious roots are found in both the horizontal and vertical rhizomes, although they are apparently absent from the vertical rhizomes of the species where these extend in the water column (e.g. *Cymodocea serrulata*, *Amphibolis* spp., and *Thalassodendron* spp.; Duarte *et al.*, 1994; Duarte *et al.*, 1998). The root epidermis of some seagrasses is lignified and several species develop sturdy roots (e.g. *Posidonia*, *Amphibolis*, and *Thalassodendron*; e.g. Kuo & McComb, 1989).

Roots often form in pairs (e.g. in Zosteraceae) associated with root-forming meristems, generally in the lower surface of rhizome nodes, leaving distinct scars after root abscission (Duarte *et al.*, 1994; Duarte *et al.*, 1998).

Some species have unbranched roots, (e.g. *Enhalus acoroides*), whereas others have densely branched root systems (e.g. *Halodule, Syringodium, Cymodocea*, Figs. 2.1, 2.2). The presence of root hairs also varies significantly among species, and are often absent in species bearing thick roots (Kuo & McComb, 1989). The meristematic area at the root tip, with a characteristic pale yellow colour, is protected by a root cap in all species. Although rich in fungi, the rhizosphere of seagrasses is apparently devoid of mycorrhiza-like structures (Nielsen *et al.*, 1999), contrary to expectations.

Seagrass roots also have a system of large lacunae in the middle cortex (Kuo & McComb, 1989), allowing the pressure-driven flow of oxygen from the photosynthetic parts to the roots (Pedersen *et al.*, 1998; Connell *et al.*, 1999). This flow is essential to maintain a supply of oxygen to support root respiration, particularly at the meristems, and to release oxygen to the rhizosphere, thereby avoiding the accumulation of phytotoxins derived from anaerobic metabolism (Hemminga, 1998). Further details on the structure of seagrass roots can be found in Kuo & McComb (1989).

2.2.4 *Flowers and fruits*

In contrast to the necessary presence of all other modules, most seagrass shoots rarely bear flowers or fruits, and the vast majority of seagrass shoots are lost before ever becoming reproductive. In general, no more than 10% of the shoots in a meadow develop flowers in any one year (e.g. Gallegos *et al.*, 1992), except for some species, such as some *Halophila* species and *Enhalus acoroides*, which can flower frequently. The latter species is able to produce up to one flower for every rhizome node formed (i.e. up to 9 flowers in a year; Duarte *et al.*, 1997a). Seagrass flowers are often inconspicuous and very simple, for they do not rely on animals for pollination. Seagrass flowers, seeds and fruits range greatly in size from the minute flowers of *Halophila* to the large male and female flowers of *Enhalus* (Fig. 2.4*a,b*).

Flowers develop from meristems positioned in the upper side of the rhizome, opposite to the root-producing meristems. Most species develop solitary flowers terminally on their vertical shoots, while some species

Fig. 2.4(*a*)

Fig. 2.4(*b*)

Fig. 2.4. Seagrass flowers and fruits: (*a*) male flowers of *Cymodocea nodosa* in south-east Spain (photograph by J. Terrados); (*b*) male *Enhalus acoroides* flower in a Philippine meadow (photograph by J. Terrados); (*c*) *Posidonia australis* flowers in Western Australia (photograph by D. Waker); (*d*) *Halophila ovalis* female flower (photograph by Japar Sidik Bujang); (*e*) fruit of *Enhalus acoroides* in a Malaysian meadow (photograph by Japar Sidik Bujang); (*f*) *Halophila spinulosa* fruits in alternate acropetal insertion (photograph by Japar Sidik Bujang).

have inflorescences (*Zostera, Posidonia, Syringodium*). *Amphibolis* and *Thalassodendron* produce viviparous seedlings, while all other species produce indehiscent fruits that are released when mature. Seagrasses often have a single fruit per plant, each containing a single seed. However, seagrass fruits are also produced in pairs in many species (*Cymodocea* spp. and *Halodule* spp.). In these species, the fruits are attached to the vertical rhizome by a short stem or peduncle, such that the fruits mature just above or within the sediments. The abscission of the flower or fruit leaves a distinct scar at the point of insertion of their peduncles onto the rhizome (Fig. 2.1; Cox & Tomlinson, 1988; Duarte *et al.*, 1994), whose width increases depending on whether the female flower developed a mature fruit or not.

Fig. 2.4(*c*)

Fig. 2.4(*d*)

2.3 Allometric scaling of seagrass form and clonal structure

The sizes of the different modules composing a seagrass ramet are not independent, but covary, such that seagrass species with thick rhizomes often have ramets characterized by large leaves and thick roots. The scaling between different plant parts is described through a series of quantitative relationships known as allometric relationships (e.g. Niklas, 1994), which are usually power equations of the form $Y = aX^b$, where Y and X are the traits whose dimensions (or mass) are compared, a is a constant correcting for the differences in magnitude between the traits, and the exponent b is the allometric coefficient, reflecting the scaling between the two dimensions compared. The values of the allometric coefficient define whether the dimensions of Y and X increase proportionally (i.e. $b = 1$), leading to isometric growth, or whether the relative dimensions (or mass) of the two properties change with size (i.e. $b \neq 1$). Module sizes are strongly correlated through allometric relationships across the seagrass flora. The rhizome diameter has been found to be an adequate parameter to scale the dimensions of other modules (Duarte, 1991; Vermaat *et al.*, 1995; Marbà & Duarte, 1998; Marbà & Walker, 1999). These scaling relationships are positive in all cases, except for internodal length, which is negativelly scaled to rhizome diameter

Fig. 2.4(*e*)

Fig. 2.4(*f*)

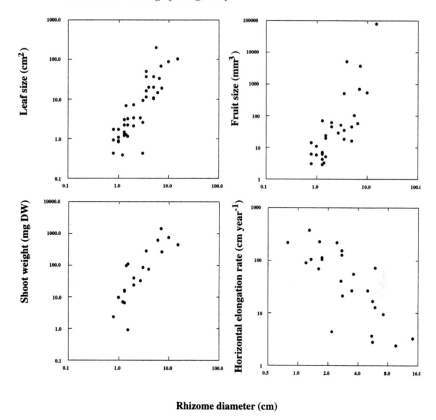

Rhizome diameter (cm)

Fig. 2.5. Allometric relationships between module sizes (leaves, fruits and shoots and rhizome elongation rate, and the diameter of seagrass rhizomes. Each data point represents an individual species. (Redrawn from Duarte, 1991.)

(Duarte, 1991). As a consequence, the horizontal elongation rate of the rhizome tends to be slower in species with thick rhizomes (Duarte, 1991; Marbà & Duarte, 1998; Fig. 2.5). The leaf surface area and the mass or volume of shoots and fruits are scaled as the square and third power of the rhizome diameter (i.e. surface L^2 and mass L^3), as expected from isometric growth (Duarte, 1991; Marbà & Duarte, 1998). These consistent scaling laws between the sizes of different modules across species imply that the repertoire of shapes in seagrasses is relatively narrow. Hence, different seagrass species can be roughly conceived as scaled models of one another, particularly when compared with the vast repertoire of forms and shapes encountered in the angiosperm flora (Bell, 1991).

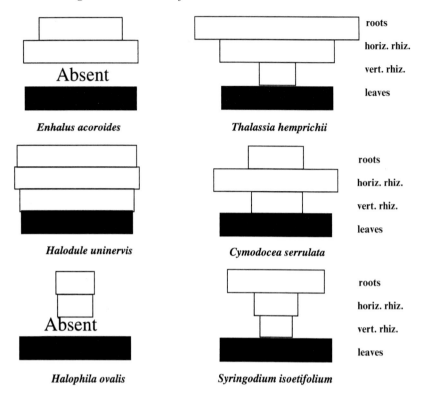

Fig. 2.6. The relative allocation of biomass to different modules in six Indo-Pacific species growing in South-East Asian stands. The length of the bars representing the weight of vertical and horizontal rhizomes and roots conforming the biomass pyramids are proportional to the weight of the leaves. (Drawn from data in Terrados *et al.*, 1999.)

The examination of the biomass allocated to the different modules of a ramet provides a parsimonious description of the basic architecture of a seagrass species. The biomass of seagrass ramets is dominated by rhizome and root material, which typically exceeds that of leaves (Fig. 2.6). The contribution of horizontal rhizomes generally exceeds that of vertical rhizomes, except for species such as *Thalassodendron* spp., which have long vertical rhizomes that extend into the water column. *Enhalus acoroides* deviates from the general pattern because its reproductive organs, such as flowers and fruits, comprise a significant fraction of the biomass of the ramets (Duarte *et al.*, 1997a), unlike the negligible fraction of biomass of the reproductive structures represented in most other species.

The number of rhizome internodes between consecutive shoots of dimorphic species varies greatly among species, with some species (e.g.

Halophila spp.) having a single internode between shoots and some others (e.g. *Thalassia* spp.) having up to 14 internodes between consecutive shoots (Duarte, 1991). Many species produce a shoot at each node, but many of these may be shed almost immediately after being produced, leading to more than one internode between standing shoots, such as in *Cymodocea* species. Accordingly, the length of rhizome between vertical rhizomes (also referred to as spacer length; Marbà & Duarte, 1998) varies greatly among species, with the vertical rhizomes of some of the species being very close together along the rhizome axis (e.g. 1.7 cm in *Halophila ovalis*), while some others are much further apart (e.g. > 6 cm in *Enhalus acoroides* or *Thalassia testudinum*; Marbà & Duarte, 1998). The spacer length is closely correlated to the size of seagrass modules, with this distance increasing as the 0.37 power of the rhizome diameter (Duarte, 1991; Marbà & Duarte, 1998). This particular scaling law has important consequences, since it determines the spacing between shoots in closed stands of the species, yielding the general decline in shoot density with increasing shoot size observed in seagrass stands, consistent with predictions from the self-thinning law (Duarte & Kalff, 1987).

The existence of close allometric relationships that constrain the architecture of seagrasses has been explained by analogy to the pipe model (cf. Marbà & Duarte, 1998), which also accounts for the relationship between stem diameter and the size of terrestrial plants (Niklas, 1994). In this analogy, the rhizome is compared to a pipe, where the flow of the substances needed to maintain clonal integration increases as the fourth power of its diameter. As a consequence, thick rhizomes should allow a better integration than thinner rhizomes do (Duarte, 1991), maintaining clonal integration for longer time periods. The hypothesized link between rhizome thickness and the persistence of clonal integration is an important factor in understanding the functioning of seagrass clones, but is yet to be quantitatively tested.

2.4 Seagrass meristems

The rate of formation of different seagrass modules, and, therefore, the growth of the clone, depends on the activity of meristems, which are the areas where active cell division takes place. Seagrasses, being monocotyledons, only have apical meristems, and lack meristematic cambium and intercalary meristems (Bell, 1991). This means that there is no secondary meristematic growth of the modules, either in diameter or length, once they are produced by the apical meristem. Hence, the full length of the leaves (and sheaths) is produced by the activity of the basal meristem. This characteristic is the basis of the leaf-marking technique commonly used to estimate seagrass leaf growth (e.g. Zieman, 1974). Seagrass growth is measured as the length (or weight) of leaf material produced between a mark on the growing leaf (usually a hole inserted with a thin needle) and a reference point (e.g. the leaf sheath) following time intervals of a few hours to days after the leaves were marked (Zieman, 1974; Kemp *et al.*, 1987). When summed over the leaves present on any given shoot, the total leaf elongation rate per shoot tends to be of the order of one or a few centimetres per shoot per day, corresponding to a turnover of the leaf material averaging 2.6% per day, although differences among species and stands can be important (Zieman, 1974; Duarte & Chiscano, 1999).

There is also some growth of newly produced modules due to cell expansion. This is limited in the leaves of most species. In *Halophila* species, however, the leaves can double in width from birth to mature size (e.g. *Halophila ovalis*, Duarte unpubl. observations). Because of the limited secondary growth, seagrass growth must be supported by continuous cell division at the apical meristems (Tomlinson, 1974), such that the growth dynamics of seagrass populations ultimately reflect the activity of their apical meristems. The apical meristem at the rhizome apex may divide at the nodes to produce branches and/or flower-producing meristems. Dimorphic seagrass species only bear flower-producing meristems at their vertical rhizomes. The formation of flower-producing or branch-producing meristems is generally preceded by an increase in the diameter of the rhizome, reflecting an increase of the meristematic area, which suggests growth conditions to be favourable.

The division of the meristems is a fairly continuous process, responsible for the maintenance and expansion of seagrass clones. The interval between the production of consecutive modules is referred to as the plastochrone interval, and is a rather conservative feature within species on an annual time scale (Table 2.2; Duarte *et al.*, 1994). The time interval between the formation of new rhizome internodes in a seagrass rhizome is also rather conservative within species, leading also to a predictable

Fig. 2.7. The relationship between the average plastochrone intervals (i.e. the time interval between the production of consecutive modules by a meristem) of the modules of different seagrass species and the average weight of shoots of different seagrass species. Each data point represents an individual species. (Redrawn from Duarte, 1991.)

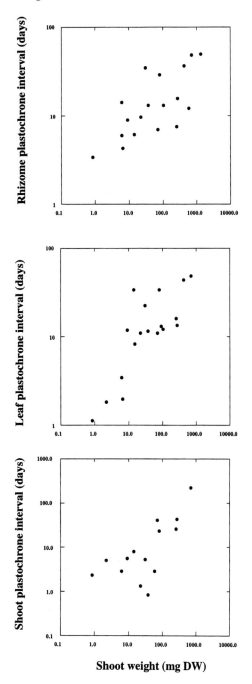

Table 2.2. *Average plastochrone intervals of the modules (leaves, horizontal rhizome internodes, and shoots) of different seagrass species*

Species	Leaf (days)	Rhizome (days)	Shoot (days)
Amphibolis antarctica	11.94	48.30	52.70
Amphibolis griffithii	31.57	87.71	
Cymodocea nodosa	31.68	28.21	22.90
Cymodocea rotundata	11.40	9.36	10.00
Cymodocea serrulata	12.70	12.69	21.20
Enhalus acoroides	35.60	35.60	
Halophila decipiens			4.50
Halophila hawaiiana	2.10	4.20	
Halophila ovalis	2.20	3.31	2.20
Heterozostera tasmanica	22.61	12.78	
Halodule uninervis	9.60	8.67	7.90
Halodule wrightii	16.52	5.88	
Posidonia angustifolia	73.59	53.33	
Posidonia australis	54.40	47.88	
Posidonia oceanica	50.68	47.81	213.00
Posidonia sinuosa	124.19	124.63	
Syringodium filiforme	60.83	33.61	5.20
Syringodium isoetifolium	33.20	6.03	7.70
Thalassia hemprichii	10.90	6.81	38.50
Thalassia testudinum	21.93	7.40	24.70
Thalassondendron ciliatum	7.89	11.83	
Thalassodendron pachyrhizum	30.81	64.19	300.00
Zostera marina	15.29	15.29	42.30
Zostera noltii	13.71	13.71	2.80

Source: Data from Duarte (1991) and Marbà & Duarte (1998).

average time span between the formation of new shoots (Table 2.2). The time span between the formation of new modules, or their plastochrone intervals (Table 2.2), is positively correlated to their size (Fig. 2.7; Duarte, 1991), such that the meristems of large seagrass species take a longer time interval to produce a new module (rhizome internode, leaf or root) than smaller species take to produce their small modules. Again, the time needed to produce a new seagrass module reflects the construction cost, which increases with size.

The activity of seagrass meristems does not proceed at constant rates or indefinitely. The rate of cell division varies seasonally, leading to changes in the rate of formation of modules, such as slower leaf production rates in winter for temperate stands (e.g. Sand-Jensen, 1975). The activity of the meristems eventually ceases. The conditions for meristematic death in seagrasses have not been studied directly, but they involve a deterioration of growth conditions, rendering the resource supply to the dividing meristems inadequate, and may involve some form of internal control mediated by hormones as well. The decline of the activity of apical meristems leads to a reduction of the meristematic area, resulting in a progressive reduction in the girth of the new internodes until the meristematic activity ceases. The death of the meristems results in the discontinuity of the production of new modules (leaves, internodes, etc.). Meristematic death is also associated with sexual reproduction in species with terminal inflorescences (e.g. *Zostera* and *Syringodium*; Kenworthy & Schwarzschild, 1998), for which reproduction, therefore, represents a terminal event. The loss of the capacity of the horizontal meristem to produce new ramets can also result from the switch to the development of a vertical rhizome, thereby producing a terminal erect shoot.

2.5 Clonal growth and branching rate: space occupation

The production of new ramets is a key component of space occupation by seagrasses, particularly during the colonization of new habitats or their recovery from disturbance (Duarte & Sand-Jensen, 1990). In addition, the production of new ramets is an important component of the production of seagrasses. For instance, the production of new shoots is reponsible for much of the production of *Zostera noltii* (Vermaat & Verhagen, 1996). Hence, clonal growth is a key trait in understanding and modelling the dynamics of seagrass populations.

2.5.1 Growth and branching of seagrass rhizomes

The growth rates of seagrass rhizomes vary greatly, from a few centimetres per year in *Posidonia oceanica* to more than 5 m yr^{-1} in *Halophila* ovalis (Duarte, 1991; Marbà & Duarte, 1998). The rate of horizontal growth of the clone is the product of the length of the internodes and their rate of formation (Duarte, 1991). As outlined above, small seagrass species have both long rhizome internodes and short plastochrone intervals, yielding, therefore, a strong negative relationship between the horizontal growth rate of seagrass rhizomes and seagrass size (Fig. 2.5). In addition, consecutive shoots of small seagrass species are produced much closer to one another than those of large species, such that the fast horizontal growth rate of small species is associated with a rapid production of new shoots (Duarte, 1991). Indeed, a meristem of a horizontal rhizome of *Halophila* may produce up to 165 shoots annually whereas that of *Posidonia oceanica* produces only 1.5 shoots per year (Table 2.2).

The vertical growth rate of seagrass rhizomes is much slower than their horizontal growth rate (Table 2.1). Vertical internodes tend to be produced at a slower rate than horizontal internodes and the vertical internodes tend to be, on average, five times shorter than the corresponding horizontal ones for any one species (Table 2.1). As a result, the rate of vertical elongation of seagrass rhizomes is about ten-fold slower than that of horizontal rhizomes (Table 2.1).

Seagrass growth is not restricted to linear growth, for seagrass rhizomes are able to branch, leading to a two-dimensional occupation of the space (Figs. 2.8 and 2.9; Brouns, 1987; Marbà & Duarte, 1998). Branching is generally monopodial, with a differentiated main axis and lateral branches, except for *Phyllospadix*, *Amphibolis*, and *Thalasso-dendron*, whose rhizomes branch sympodially. Monopodial branching is generally dichotomic, with branches produced on alternate sides of the main axis. Indeed, the vertical rhizomes of many seagrass species (e.g. *Posidonia* spp., *Enhalus acoroides*) arise as lateral branches, and differentiate as vertical shoots subsequently. Average branching rates of seagrass rhizomes range widely from one branch for every four nodes in *Heterozostera tasmanica* to one branch in 1600 internodes in *Thalassia testudinum*. Small seagrass species tend to branch much more profusively than large seagrass species do (Marbà & Duarte, 1998). Hence, small seagrass species elongate faster and spread in two dimensions at a greater rate than large seagrass species do (Fig. 2.9). It is, therefore, not surprising that small seagrass species are considered to play a pioneer role in multispecific floras (cf. Duarte, 1991).

Branching in seagrasses is a tightly regulated process, involving a suppression of branching in the proximity of actively growing rhizome apices. This process, referred to as 'apical dominance' is widespread in rhizomatous plants, and serves as a mechanism providing the capacity to shift the patterns and direction of clonal growth. The occurrence of apical dominance in seagrasses is, however, poorly documented, with most of the evidence consisting of observations of increased branching rate following damage of apical meristems and the experimental supression of apical meristems in the horizontal rhizomes of *Cymodocea nodosa* (Terrados *et al.*, 1997a,b). The finding that branching increases following the manipulation of hormone levels in seagrasses (Terrados-Muñoz, 1995) suggests that the suppression of branching through apical dominance involves hormonal regulation. The distance over which apical

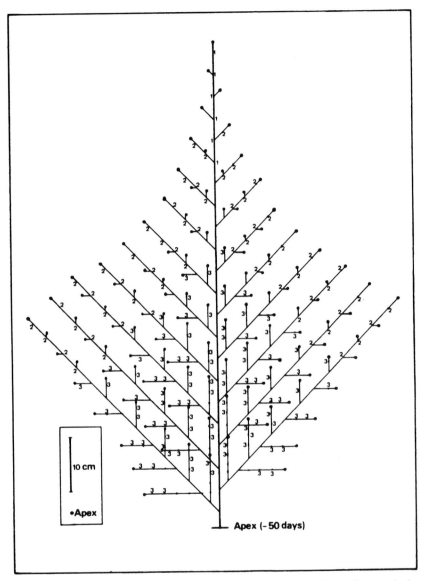

Fig. 2.8. Topology of a clone formed over 50 days of growth by *Halophila ovalis*. 1, leaf pairs on the main axis present; 2, leaf pairs on the axes of the second order present; 3, leaf pairs on the axes of the third order present. (Brouns, 1987.)

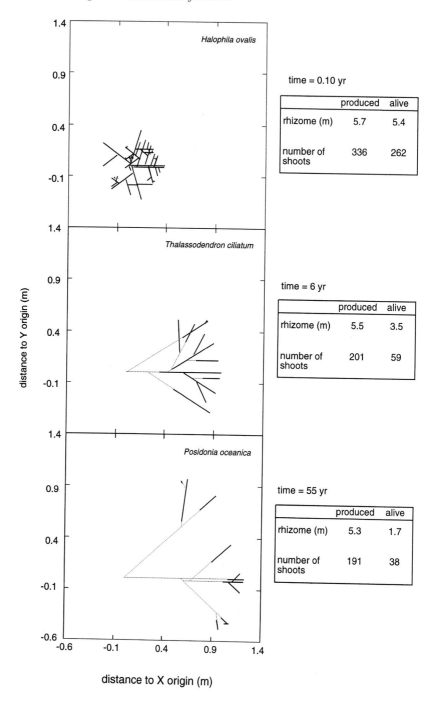

dominance is effective in seagrasses has been experimentally determined to be in the order of 0.5 m for *Cymodocea* nodosa (Terrados *et al.*, 1997a), but it is unknown for any other seagrass species. The much lower branching rate of large seagrass species compared with small ones suggests that the length over which apical dominance is effective increases with increasing seagrass size.

The geometry of seagrass clonal networks and their efficiency in occupying space, as defined by the largest surface covered with the minimum investment in rhizome material, is critically dependent on the angle at which the horizontal rhizomes branch (Marbà & Duarte, 1998). The dependence of the efficiency of two-dimensional occupation on the branching angle has been demonstrated for all kinds of branching systems (Leopold, 1971; Stevens, 1974) and explicitly investigated for clonal plants (Bell & Tomlinson, 1980). Geometric network analyses have shown that branching angles of 30 degrees provide the most efficient occupation of space (Stevens, 1974), leading to the greatest occupation of space with the minimum investment in rhizome material. Indeed, most seagrass species branch at angles of about 40 degrees, ranging from 19 to 81 degrees (Marbà & Duarte, 1998). The branching angle increases significantly with increasing seagrass size, such that the smallest seagrass species (e.g. *Halophila*) have branching angles of about 50–80 degrees (Marbà & Duarte, 1998). The dependence of branching angle on rhizome diameter, similar to that observed in other branching systems, is explained by the 'minimal effort' principle, by which branched networks are designed to minimize the energy needed to transport materials (Leopold, 1971; Stevens, 1974; Vogel 1981). The examination of the distribution of branching angles leads to the prediction that large, slow-growing seagrass species should be able to achieve the greatest spatial coverage with the minimum investment in rhizomes, whereas small

Fig. 2.9. The simulated spread of clones of different seagrass species predicted on the basis of their basic growth rules: horizontal rhizome elongation rate, and branching rules (probability and angle). The graphs depict the clonal topography after producing ca. 5 m of rhizome for three contrasting seagrass species (*Halophila ovalis*, *Thalassodendron ciliatum*, *Posidonia oceanica*). The time required to develop the networks, and the rhizome length and number of shoots produced and surviving since initiation of clonal spread are indicated. Dashed lines show the spatial distribution of the rhizomes and shoots produced, and continuous lines the distribution of surviving rhizomes and shoots. (Marbà & Duarte, 1998.)

seagrass species should have a less efficient, but more compact, occupation of the space (Marbà & Duarte, 1998).

The basic components of the clonal spreading of seagrasses are, thus, the rate of horizontal extension, and the probability and angle of branching of the rhizomes. The clonal growth of seagrasses can, therefore, be simulated from knowledge of this basic set of rules, an approach that has proved most useful in the examination of space occupation by clonal plants (Bell & Tomlinson, 1980). The simulation of the space occupation by seagrass clones confirms the prediction that small seagrass species have a less efficient but more compact occupation of space, which has been referred to as the 'phalanx' strategy, whereas large, slow-growing species have a more efficient, but looser, occupation of the space, the 'guerrilla strategy' (Fig. 2.9). If large species did have broader branching angles, the time required to occupy the space would be so long that they would not be able to develop meadows despite the long life span of their modules (Marbà & Duarte, 1998). A simulation analysis of seagrass clonal growth also showed that the branching process continuously accelerates the occupation of space, such that the space occupied by a seagrass clone increases as the third power of time for all seagrass species simulated. Indeed, branching rates and branching angles are even more important determinants of the rate of space occupation than the linear extension rate of the rhizomes, the parameter that has received most attention to date (Marbà & Duarte, 1998).

2.6 Adaptive value of seagrass architecture and clonal growth

The architecture of seagrasses has been presented hitherto as a static property of the species, whereas, in fact, plasticity in the architecture of seagrasses is substantial, and an important component of their capacity to adapt to disturbance or resource heterogeneity. For instance, horizontal rhizomes have an exploratory capacity, being responsible for the extension of the clone into new areas, while vertical rhizomes have the capacity to use resources, deploying leaves and roots at the same position to use the local resources. Changes in architecture and clonal growth may, thus, influence both the exploratory and the resource use potential of the clone.

2.6.1 Plasticity of seagrass modules

The extent of intraspecific plasticity of seagrass modules has been quantified as the coefficient of variation of the documented distribution of module size within species (Duarte, 1991; Marbà & Duarte, 1998; Terrados *et al.*, 1999). This examination has shown the size of seagrass modules to be very plastic, except for the number of leaves in each shoot, and, particularly, the diameter of seagrass rhizomes, which are the most conservative traits within individual species (Duarte, 1991; Marbà & Duarte, 1998; Terrados *et al.*, 1999). Indeed, the rhizome diameter and appearance are so conservative within species that we suggest that it should be explored as possibly more reliable for diagnostic purposes in seagrass taxonomy than other more plastic traits (e.g. venation and shape of leaf apex) that are presently used.

The sequence of addition of modules, or their plastochrone intervals, which reflects the internal clocks of seagrass meristems, is also relatively conservative within species (coefficient of variation ca. 30% within species; Duarte, 1991), although their relative variability differs across species. The rate of addition of modules is extremely conservative in *Cymodocea nodosa* (mean = 12 leaves shoot^{-1} year^{-1}, coefficient of variation = 10% among 22 populations; Duarte *et al.*, unpubl. results), whereas it shows great variability in other species (e.g. *Posidonia oceanica*; Pergent & Pergent-Martini, 1990). The considerable variability in the length of both vertical and horizontal internodes, together with the smaller variability in the rate of production of new internodes and vertical rhizomes, results in highly variable elongation rates of seagrass rhizomes within individual species.

Variability in module size or formation rate often has a seasonal pattern, particularly in temperate climates, with small leaves and rhizome internodes being produced at low rates during the adverse season (e.g. Cayé & Meinesz, 1985; Duarte & Sand-Jensen, 1990). The rate of formation of new shoots also appears to decline during the adverse season (e.g. Cayé & Meinesz, 1985). Seasonality in module size and formation has also been reported in tropical and subtropical climates, where it may be triggered by intrinsic events (e.g. sexual reproduction) or extrinsic factors, such as tidal cycles.

The considerable plasticity in module size discussed above suggests that it may have an important adaptive value, and may be, therefore, an important component of the capacity of seagrasses to cope with changes and heterogeneity in the environment. In turn, the plasticity of seagrass modules may also respond to the availability of resources, reducing the construction costs when resources are scarce. Changes in shoot and leaf size affect the capacity to acquire light. Hence, there appears to be a general tendency for leaf and shoot size to increase with increasing depth (e.g. Dennison & Alberte, 1986). Conversely, leaf and shoot size tend to be small when resources other than light are limiting, such as in nutrient-limited stands (Pérez *et al.*, 1994; Agawin *et al.*, 1996; Terrados *et al.*, 1999).

The considerable plasticity of root length and branching also has obvious adaptive value. The root systems of nutrient-limited stands are often highly branched (Pérez *et al.*, 1994), thereby increasing the absorptive root surface. In contrast, the root systems of stands growing in nutrient-rich environments are often composed of a single unbranched root, which typically is long and thick. The development of thick, long unbranched roots in nutrient-rich sediments may lead to a possible greater anchoring capacity (Bell, 1991) or efficiency in oxygen transport of such root systems, although the functional implications of this root form remain to be tested. In fact, the absolute biomass allocation to roots appears to be relatively uniform, and the plasticity of root systems results from an increase in their number or branching and a reduction in their length in nutrient-poor habitats (e.g. Terrados *et al.*, 1999). However, the relative biomass allocation to root systems increases in nutrient-poor habitats, where shoot size, but not root mass, is reduced (Pérez *et al.*, 1994, Terrados *et al.*, 1999).

The considerable plasticity in horizontal internodal length has important consequences for the occupation of space. The internodal length tends to be greater in deep-growing stands, thereby achieving a greater separation between shoots, decreasing self-shading. Similarly, the internodal length appears to be greater in colonizing rhizome axes, allowing, therefore, a fast occupation of the space, whereas the internodal length is reduced in crowded seagrass stands. The empirical evidence supporting these tendencies is, however, still insufficient to assess whether they reflect general trends or particular responses of the species so far investigated. In principle, the occupation of space by seagrass rhizomes can be modulated by changes in the rate of spread (largely derived from changes in internodal length), but also in the extent of exploitation of the space, by changes in the frequency of branching. It has been proposed that such changes enable clonal plants to display a foraging behaviour in spatially heterogeneous environments, whereby plants develop a greater density of ramets in resource-rich patches (Cain *et al.*, 1996). The existence of such a foraging behaviour, although possible in theory, has been tested for only a few seagrasses (Tomasko, 1992). Such behaviour is, however, often observed in seagrass roots, which are often found to penetrate inside empty mollusc shells and dead rhizomes in the sediment, where enhanced nutrient concentrations are expected.

The adaptive value of plasticity in vertical rhizome internodes has been explored well beyond that of any other module. Vertical rhizomes raise the leaf-producing meristem above the level of the horizontal rhizome, allowing the meristems to be raised closer to the sediment surface when burial occurs, and to be raised higher above the sediment surface in lush meadows where bottom-dwelling plants may be strongly self-shaded. The increase in the length of vertical internodes of buried seagrass stands has been reported for a number of species in the Caribbean (Patriquin, 1973; Marbà *et al.*, 1994a), the Mediterranean (Marbà & Duarte, 1994; Marbà *et al.*, 1994b), and South-East Asia (Duarte *et al.*, 1997b). Following burial, the internodal length of vertical rhizomes may increase greatly relative to the average value. Concurrently, the vertical elongation rate of buried shoots increases proportionally, allowing the plants to survive disturbances causing strong sedimentation. For instance, the motion of mega sand waves following Hurricane Gilbert buried stands of *Thalassia testudinum* under more than a metre of sand in Bahía Mujeres (Yucatán, Mexico; Marbà *et al.*, 1994a). Some shoots survived by greatly increasing the length of their vertical internodes to raise the leaves to the sediment surface. Increased internodal length also allowed *Cymodocea nodosa* shoots to survive the passage of sand waves in the Spanish Mediterranean (Marbà *et al.*, 1994b). The enhanced elongation of vertical rhizomes is maintained until the leaf-producing meristem is repositioned at the sediment surface, and the vertical growth rate resumes its regular average rate thereafter (Marbà *et al.*, 1994b; Duarte *et al.*, 1997b). This behaviour suggests that the increased elongation of buried vertical rhizomes is triggered by the absence of light. This was confirmed experimentally by the observation that the response was suppressed in buried shoots whose meristems were illuminated within the sediments using fibre optics (Terrados, 1997). Increased vertical growth of buried shoots is often accompanied by high shoot mortality, suggesting that resources reabsorbed from dead shoots may help maintain the fast vertical growth rates of the surviving shoots. In addition, burial of seagrass rhizomes often leads to the mortality of horizontal rhizomes (Marbà & Duarte, 1995), which, if growth continued, would produce new shoots buried deep in the sediments that would burden the carbon balance of the plant. The apical meristem of fast-growing vertical rhizomes of buried seagrasses often branches once it is close to the sediment surface, yielding a horizontal rhizome that will extend over the new sediment horizon (Marbà & Duarte, 1995; Duarte *et al.*, 1997b).

In many seagrass species (e.g. *Thalassodendron* and *Amphibolis*) the vertical rhizomes can be raised above the sediment surface, thereby positioning the leaves higher in the water column. In this way shading by other plants present in dense stands is avoided. The length of the vertical rhizome internodes of some of these species is also greatest when the shoots are still young, and declines as the leaf-producing meristem is raised further above the sediment surface with increasing age (e.g. *Thalassodendron*; Duarte *et al.*, 1996). The capacity to extend vertically into the water column requires special adaptations in the rhizomes, such as an increased lignification to maintain the vertical position outside the sediments and the capacity to avoid the effect of drilling organisms, whose activity leads to the death of exposed seagrass rhizomes in other species (Marbà & Duarte, 1995). In addition to the species indicated above, whose vertical rhizomes always extend above the sediment surface, some species whose vertical rhizomes usually remain inside the sediments may also extend them into the water column in dense stands (e.g. *Halodule wrightii*; Fourqurean *et al.*, 1995; *Halodule uninervis*; Duarte, unpubl. data; *Syringodium filiforme*; Kenworthy & Schwarzchild, 1998). Branching of the raised rhizomes of these species has been reported to produce multilayered, 3-D canopies (Kenworthy & Schwarzschild, 1998). This phenomenon has been observed in nutrient-rich waters, and may be a response to increased competition for light or deterioration of sediment conditions associated with eutrophication (see section 7.5). Such responses have a cost, in that the path length between neighbouring meristems is increased due to the longer rhizomes, thereby increasing the resistence for translocation and clonal integration. In addition, the respiratory tissue is increased relative to the photosynthetic tissue, increasing overall carbon demands (Kenworthy & Schwarzschild, 1998).

2.7 Concluding remarks

Seagrasses are modular plants sharing a common architecture with many terrestrial and freshwater grasses, so that their evolutionary history in the sea has contributed little variation to the architecture of monocots. The repertoire of seagrass form and shape is surprisingly narrow, such that much of the variation among species can be explained by simple scaling laws. This narrow range of variability suggests these architectural features have an important impact on fitness. Indeed, the narrow plasticity in architecture present within species has important adaptive value, allowing the plants to accommodate to varying resource levels and to modulate the rates at which they occupy the space. As will be discussed in the next chapter, the population dynamics of seagrasses represents, to a large extent, a projection of the basic growth rules controlling their architecture to the much larger spatial scales occupied by seagrass clones and meadows.

2.8 References

Agawin, N.S.R., Duarte, C.M. & Fortes, M.D. (1996). Nutrient limitation of Philippine seagrasses (Cape Bolinao, NW Philippines): *in situ* experimental evidence. *Marine Ecology Progress Series*, **138**, 233–43.

Agustí, S., Enríquez, S., Christensen, H., Sand-Jensen, K. & Duarte, C.M. (1994). Light harvesting among photosynthetic organisms. *Functional Ecology*, **8**, 273–9.

Bell, A.D. (1991). *Plant Form*. New York: Oxford University Press.

Bell, A.D. & Tomlinson, P.B. (1980). Adaptive architecture in rhizomatous plants. *Botanical Journal of the Linnean Society*, **80**, 125–60.

Brouns, J.W.M. (1987). Growth patterns in some indo-West-Pacific seagrasses. *Aquatic Botany*, **28**, 39–61.

Cain, M.L., Dudle, K.A. & Evans, J. P. (1996). Spatial models of foraging in clonal plant species. *American Journal of Botany*, **83**, 76–85.

Cayé, G. & Meinesz, A. (1985). Observations on the vegetative development, flowering and seedling of *Cymodocea nodosa* (Ucria) Ascherson on the Mediterranean coasts of France. *Aquatic Botany*, **22**, 277–89.

Connell, E.L., Colmer, T.D. & Walker, D.I. (1999). Radial oxygen loss from intact roots of *Halophila ovalis* as a function of distance behind the root tip and shoot illumination. *Aquatic Botany*, **63**, 219–28.

Cox, P.A. & Tomlinson, P.B. (1988). Pollination ecology of a seagrass, *Thalassia testudinum* (Hydrocharitaceae) in St. Croix. *American Journal of Botany*, **75**, 958–65.

Den Hartog, C. (1970). *The Seagrasses of the World*. Amsterdam: New Holland Publishers.

Dennison, W.C. & Alberte, R.S. (1986). Photoadaptation and growth of *Zostera marina* L. (eelgrass) transplants along a depth gradient. *Journal of Experimental Marine Biology and Ecology*, **98**, 265–82.

Duarte, C.M. (1991). Allometric scaling of seagrass form and productivity. *Marine Ecology Progress Series*, **77**, 289–300.

Duarte, C.M. & Chiscano, C.L. (1999). Seagrass biomass and production: a reassessment. *Aquatic Botany*, **65**, 159–74.

Duarte, C.M. & Kalff, J. (1987). Weight-density relationships in submerged macrophytes: the importance of light and plant geometry. *Oecologia (Berlin)*, **72**, 612–17.

Duarte, C.M. & Sand-Jensen, K. (1990). Seagrass colonization: patch formation and patch growth in *Cymodocea nodosa*. *Marine Ecology Progress Series*, **65**, 193–200.

62   *Seagrass architectural features*

Duarte, C.M., Marbà, N., Agawin, N., Cebrián, J., Enríquez, S., Fortes, M.D., Gallegos, M.E., Merino, M., Olesen, B., Sand-Jensen, K., Uri, J. & Vermaat, J. (1994). Reconstruction of seagrass dynamics: age determinations and associated tools for the seagrass ecologist. *Marine Ecology Progress Series*, **107**, 195–209.

Duarte, C.M., Hemminga, M. & Marbà, N. (1996). Growth and population dynamics of *Thalassodendron ciliatum* in a Kenyan back-reef lagoon. *Aquatic Botany*, **55**, 1–11.

Duarte, C.M., Uri, J., Agawin, N.S.R., Fortes, M.D., Vermaat, J.E. & Marbà, N. (1997a) Flowering frequency of Philippine seagrasses. *Botanica Marina*, **40**, 497–500.

Duarte, C.M., Terrados, J., Agawin, N.S.W., Fortes, M.D., Bach, S., & Kenworthy, W.J. (1997b). Response of a mixed Philippine seagrass meadow to experimental burial. *Marine Ecology Progress Series*, **147**, 285–94.

Duarte, C.M., Merino, M., Agawin, N.S.R., Uri, J., Fortes, M.D., Gallegos, M.E., Marbà, N. & Hemminga, M. (1998). Root production and belowground seagrass biomass. *Marine Ecology Progress Series*, **171**, 97–108.

Enríquez, S., Agustí, S. & Duarte, C.M. (1992). Light absorption by seagrass (*Posidonia oceanica* (L.) Delile) leaves. *Marine Ecology Progress Series*, **86**, 201–4.

Enríquez, S., Duarte, C.M. & Sand-Jensen, K. (1993). Patterns in decomposition rates among photosynthetic organisms: the importance of detritus C:N:P content. *Oecologia*, **94**, 457–71.

Fourqurean, J.W., Powell, G.V.N., Kenworthy, W.J. & Zieman, J.C. (1995). The effects of long-term manipulation of nutrient supply on competition between the seagrasses *Thalassia testudinum* and *Halodule wrightii* in Florida Bay. *Oikos*, **72**, 349–58.

Gallegos, M.E., Marbà, N., Merino, M. & Duarte, C.M. (1992). Flowering of *Thalassia testudinum* Banks ex König in the Mexican Caribbean: age-dependence and interannual variability. *Aquatic Botany*, **43**, 249–55.

Grice, A.M., Loneragan, N.R. & Dennison, W.C. (1996). Light intensity and the interaction between physiology, morphology and stable isotope ratios in five species of seagrass. *Journal of Experimental Marine Biology and Ecology*, **195**, 91–110.

Hemminga, M.A. (1998). The root/rhizome system of seagrasses: an asset and a burden. *Journal of Sea Research*, **39**, 183–196.

Kemp, W.M., Murray, L., Borum, J. & Sand-Jensen, K. (1987). Diel growth in eelgrass *Zostera marina*. *Marine Ecology Progress Series*, **41**, 79–86.

Kenworthy, W.J. & Schwarzschild, A.C. (1998). Vertical growth and short-shoot demography of *Syringodium filiforme* in outer Florida Bay, USA. *Marine Ecology Progress Series*, **173**, 25–37.

Klap, V.A., Hemminga, M.A. & Boon, J.J. (2000). The retention of lignin in seagrasses, angiosperms that returned to the sea. *Marine Ecology Progress Series* (in press).

Kuo, J. & McComb, A.J. (1989). Seagrass taxonomy, structure and development. In *Biology of Seagrasses*, ed. A.W.D. Larkum, A.J. McComb, & S.A. Sheperd, pp. 6–73. Amsterdam: Elsevier.

Leopold, L.B. (1971). Trees and streams: the efficiency of branching patterns. *Journal of Theoretical Biology*, **31**, 339–54.

Marbà, N. & Duarte, C.M. (1994). Growth response of the seagrass *Cymodocea nodosa* to experimental burial and erosion. *Marine Ecology Progress Series*, **107**, 307–11.

Marbà, N. & Duarte, C.M. (1995). Coupling of seagrass (*Cymodocea nodosa*) patch dynamics to subaqueous dune migration. *Journal of Ecology*, **83**, 381–9.

Marbà, N. & Duarte, C.M. (1998). Rhizome elongation and seagrass clonal growth. *Marine Ecology Progress Series*, **174**, 269–80.

Marbà, N. & Walker, D. I. (1999). Population dynamics of temperate Western Australian seagrasses: importance of growth and flowering for meadow maintenance. *Marine Ecology Progress Series*, **184**, 105–18.

Marbà N., Gallegos, M.E., Merino, M. & Duarte, C.M. (1994a). Vertical growth of *Thalassia testudinum*: seasonal and interannual variability. *Aquatic Botany*, **47**, 1–11.

Marbà, N., Cebrián, J., Enríquez, S. & Duarte, C.M. (1994b). Migration of large-scale subaqueous bedforms measured with seagrasses (*Cymodocea nodosa*) as tracers. *Limnology and Oceanography*, **39**, 126–33.

Nielsen, S.L., Thingstrup, I. & Wigand, C. (1999). Apparent lack of vesicular-arbuscular mycorrhiza (VAM) in the seagrasses *Zostera marina* L. and *Thalassia testudinum* Banks ex König. *Aquatic Botany*, **63**, 261–6.

Niklas, K.J. (1992). *Plant Biomechanics: an engineering approach to plant form and function*. Chicago: University of Chicago Press.

Niklas, K.J. (1994). *Plant Allometry: the scaling of form and process*. Chicago: University of Chicago Press.

Patriquin, D.G. (1973). Estimation of growth rate, production and age of the marine angiosperm *Thalassia testudinum* König. *Caribbean Journal of Science*, **13**, 111–23.

Patriquin, D.G. (1975). 'Migration' of blowouts in seagrass beds at Barbados and Carriacou, West Indies, and its ecological and geological implications. *Aquatic Botany*, **1**, 163–89.

Pedersen, O., Borum, J., Duarte, C.M. & Fortes, M.D. (1998). Oxygen dynamics in the rhizosphere of *Cymodocea rotundata*. *Marine Ecology Progress Series*, **169**, 283–8.

Pérez, M., Duarte, C.M., Romero, J., Sand-Jensen, K. & Alcoverro, T. (1994). Growth plasticity in *Cymodocea nodosa* stands: the importance of nutrient supply. *Aquatic Botany*, **47**, 249–64.

Pergent, G. & Pergent-Martini, G. (1990). Some applications of lepidochronological analysis in the seagrass *Posidonia oceanica*. *Botanica Marina*, **33**, 299–310.

Roberts, D.G. & Moriarty, D.J.W. (1987). Lacunal gas discharge as a measure of productivity in the seagrasses *Zostera capricorni*, *Cymodocea serrulata* and *Syringodium isoetifolium*. *Aquatic Botany*, **28**, 143–60.

Sand-Jensen, K. (1975). Biomass, net production and growth dynamics in an eelgrass (*Zostera marina* L.) population in Vellerup Vig, Denmark. *Ophelia*, **14**, 185–201.

Sculthorpe, C.D. (1967). *The Biology of Aquatic Vascular Plants*. London: Arnold.

Stevens, P.S. (1974). *Patterns in Nature*. Boston: Atlantic Monthly Press.

Terrados, J. (1997). Is light involved in the vertical growth response of seagrasses when buried by sand? *Marine Ecology Progress Series*, **152**, 295–9.

Terrados, J., Duarte, C.M. & Kenworthy, W.J. (1997a) Is the growth of the apical meristem of seagrasses dependent on clonal integration? *Marine Ecology Progress Series*, **158**, 103–10.

Terrados, J., Duarte, C.M. & Kenworthy, W.J. (1997b) Experimental evidence for apical dominance in the seagrass *Cymodocea nodosa* (Ucria) Ascherson *Marine Ecology Progress Series*, **147**, 263–8.

Terrados, J., Borum, J., Duarte, C.M., Fortes, M.D., Kamp-Nielsen, L., Agawin, N.S.R. & Kenworthy, W.J. (1999). Nutrient and biomass allocation of SE Asian seagrasses. *Aquatic Botany*, **63**, 203–17.

Terrados-Muñoz, J. (1995). Effects of some plant growth regulators on the growth of the seagrass *Cymodocea nodosa* (Ucria) Ascherson. *Aquatic Botany*, **51**, 311–18.

Tomasko, D.A. (1992). Variation in growth form of shoal grass (*Halodule wrigthtii*) due to changes in the spectral composition of light below a canopy of turtle grass (*Thalassia testudinum*). *Estuaries*, **15**, 214–17.

Tomlinson, P.B. (1974). Vegetative morphology and meristem dependence. The foundation of productivity in seagrasses. *Aquaculture*, **4**, 107–30.

Vermaat, J.E. & Verhagen, F.C.A. (1996). Seasonal variation in the intertidal seagrass *Zostera noltii* Hornem: coupling demographic and physiological patterns. *Aquatic Botany*, **52**, 259–81.

Vermaat, J.E., Fortes, M.D., Agawin, N.S.R., Duarte, C.M., Marbà, N. & Uri, J. (1995). Meadow maintenance, growth, and productivity of a mixed Philippine seagrass bed. *Marine Ecology Progress Series*, **124**, 215–55.

Vermaat, J.E., Agawin, N.S.R., Fortes, M.D., Uri, J.S., Duarte, C.M., Marbà, N., Enríquez, S. & van Vierssen, W. (1997). The capacity of seagrasses to survive increased turbidity and siltation: the significance of growth form and light use. *Ambio*, **26**, 499–504.

Vogel, S. (1981). *Life in Moving Fluids*. Princeton: Princeton University Press.

Zieman, J.C. (1974). Methods for the study of the growth and production of the turtle grass, *Thalassia testudinum* König. *Aquaculture*, **4**, 139–43.

3

Population and community dynamics

3.1 Introduction

Seagrass meadows often appear to the casual observer as static land-scapes. However, seagrass meadows are subject to intense dynamics involving the continuous loss and replacement of shoots in the population, which, when in balance, maintain a dynamic equilibrium. Such apparent steady-state conditions can be maintained over extended time periods, leading to long-lived seagrass meadows, such as some *Posidonia oceanica* meadows, which possibly may persist for > 4 000 years in the Mediterranean (e.g. Mateo *et al.*, 1997), and *Zostera marina* meadows exceeding a millennium in age (Reusch *et al.*, 1999).

The equilibrium maintaining seagrass meadows is, however, often upset, leading to a regression of seagrass meadows, whereby large meadows can be totally lost over a few years (see Chapter 7). In fact, seagrass decline is now a common phenomenon throughout the world (Short & Wyllie-Echeverria, 1996), to the point that the law in various countries now protects seagrass meadows. However, effective protection of seagrass meadows requires an understanding of the regulation of seagrass losses and gains. This is currently the bottleneck to the development of reliable forecasts on the future status of seagrass meadows. Examination of the dynamics of genetic individuals (genets) and patches within the meadow requires knowledge of the life cycles of seagrasses. Seagrass life cycles are similar to those of clonal herbs on land, except that they are entirely confined to the marine realm. However, dispersal processes are remarkably different in the underwater marine environment compared with land.

Study of the population dynamics of seagrasses is complicated by their clonal nature and the difficulties in discriminating genets within a population, which is only possible with the use of molecular techniques. Moreover, the population dynamics of seagrasses should be approached in a hierarchical manner, addressing the demography (fluctuations in abundance) of individual modules (e.g. leaves), ramets (shoots), genets (clones) and landscapes (patches and meadows). In practice, the dynamics of the population of shoots gets much of the attention.

Although temperate seagrass meadows are typically monospecific, tropical meadows can be highly diverse, containing up to 11 species. The maintenance of these meadows does not depend only on the demography and the response to growth conditions of the individual species, but it also depends on competitive interactions and positive interactions with other species in the community. The dynamics of the community has been examined by following the succession process after perturbations and, in some cases, through experimental approaches, largely addressing competitive interactions between species. Both the observational studies and competition experiments emphasize the interactions resulting in species replacements, and cannot account for the coexistence of species in mixed meadows. The maintenance of mixed meadows requires either perturbations maintaining the communities in dynamic equilibrium or positive interactions to counterbalance the possible negative, competitive interactions.

In this chapter we address seagrass population and community dynamics at increasing levels of organization. As an introduction to the subject, we provide a discussion on the interpretation and reconstruction of seagrass growth history.

3.2 Interpreting seagrass growth history

A basic understanding of seagrass architecture and meristem dynamics (Chapter 2) allows the interpretation of seagrass growth from examination of the sequence of their modules. Nodes indicate the insertion line of leaves, whether functional or bracteae (i.e. greatly reduced leaves). The leaves are produced on the rhizome in two rows on opposite sides (distichous phyllotaxis). Hence, consecutive leaves open to alternative sides of the shoot, resulting in an alternation of nodes along the rhizome axis in species with open nodes. The reconstruction of leaf age ranks (i.e. the rank order, from youngest to oldest, of the leaves present in a shoot) is, therefore, very simple; younger leaves are enclosed by older leaves. The age rank of leaves can be translated into absolute time whenever knowledge of the time interval between production of consecutive leaves (i.e. the leaf plastochrone interval) is available. The product of the leaf plastochrone interval and the rank position of the leaves provides, when applying adequate corrections to account for the fraction of a plastochrone interval lived by the youngest leaf, a reliable estimate of leaf age (Cebrián *et al.*, 1994). The capacity to derive the age of seagrass leaves has simple direct applications, such as the calculation of grazing rates, from the relationship between the leaf surface area lost by grazing activity and leaf age (Cebrián *et al.*, 1994). A rough estimate of the life span of leaves can be derived as the product of the mean number of leaves per shoot and their average plastochrone interval. The applications associated with knowledge of the average leaf plastochrone interval extend beyond the calculation of leaf age as they allow the estimation of the age of other seagrass modules, providing the basis of demographic growth analysis (Duarte *et al.*, 1994).

The leaf plastochrone interval is believed to respond to internal clocks in higher plants, representing the time units for their biological age (Lamoreaux *et al.*, 1978). As indicated in the previous chapter, the average number of leaves produced annually on a seagrass shoot can be very uniform across populations of some species but can show substantial variability in others (Table 2.2). Interannual variation within populations has also been documented (e.g. Marbà & Duarte, 1997), but this is often relatively small compared with the variation among populations. Hence, average estimates of the duration of the plastochrone interval for a given seagrass species provide only rough approximations, and should be empirically established for the populations studied. This can be accomplished using leaf marking techniques, whereby the leaves on the shoots are marked, typically by punching a hole through their basis with a needle, and the number of new – unpunched – leaves formed after an adequate time interval is determined (Brouns, 1985). It should be added that the average leaf plastochrone interval of leaves produced on the horizontal rhizomes is often shorter than that of leaves produced on vertical rhizomes in seagrasses with dimorphic rhizomes (cf. Table 2.2).

The rate of rhizome growth is difficult to estimate using marking techniques, particularly for species with rhizomes buried deep down into the sediments (e.g. *Enhalus acoroides*, *Posidonia oceanica*, *Thalassia testudinum*; Duarte *et al.*, 1998). The presence of leaf scars left by shed leaves allows the estimation of the age of seagrass shoots, as the product of the total number of leaf scars present on the shoot (i.e. the number of leaves produced over their life span) and the average leaf plastochrone interval (Patriquin, 1973). The ability to estimate the age of seagrass

Fig. 3.1. An apical rhizome of *Thalassodendron ciliatum* (Kenya) showing the gradation from very young, small, shoots near the rhizome apex, to older, larger shoots at increasing distances from the apex. The loss of meristematic activity in some of the shoots (the leafless shoots) leads to their death. (Photograph by N. Marbà.)

shoots or rhizomes also provides a simple way to estimate rhizome growth. The ratio between the length of rhizome between distant shoots and the age difference between them (Fig. 3.1), representing the time span over which the rhizome piece was formed (Duarte *et al.*, 1994), provides an estimate of the rate of horizontal extension of seagrasses (cf. Patriquin, 1973; Marbà & Duarte, 1998). Similarly, the rate of vertical extension of seagrass rhizomes can be estimated as the ratio between the length of vertical rhizome and shoot age (Duarte *et al.*, 1994). The rate of vertical extension of seagrass rhizomes ranges from absent in species lacking vertical rhizomes to a maximum of 13 cm year^{-1} reported for some populations of *Cymodocea serrulata* (Vermaat *et al.*, 1995) and *Syringodium filiforme* (Kenworthy & Schwarzschild, 1998).

Leaf scars are not the only remnants of lost modules in seagrass rhizomes; all lost modules (roots, branches, shoots, flower, fruits, and leaves) leave scars on the rhizomes when they are shed (Chapter 2). The scars left by the abscission of modules are regular in shape and location, and can, therefore, be identified easily (Bell, 1991). The position of scars of different types on rhizomes aids in the interpretation of seagrass growth history, as well as in the identification of key events, such as flowering.

A temporary cessation of growth, generally associated with adverse seasons for growth, is often associated with the presence of very short internodes and crowded leaf scars (Bell, 1991), indicative of a slow rhizome growth. Hence, winter growth is often identified in temperate seagrasses by rhizome portions with short internodes (e.g. Cayé & Meinesz, 1985; Duarte & Sand-Jensen, 1990a). An increased spacing of leaf scars along the vertical rhizomes, resulting from enhanced vertical growth, is indicative of sediment burial (Fig. 3.2; e.g. Marbà *et al.*, 1994; Duarte *et al.*, 1997). Hence, leaf scars and internodes of seagrass rhizomes provide a record of past growth and key events, which can help reconstruct their past history, as well as sedimentary events, such as the formation of blow-outs during storms (Patriquin, 1975) and the motion of sand waves (Marbà *et al.*, 1994a,b); which are difficult to observe directly.

3.3 Life span of seagrass modules

Meristematic death is followed by the loss of functionality of the modules, which may subsequently be shed, thereby avoiding respiratory losses by non-functional organs. For example, leaves are shed shortly after their basal meristem stops dividing. The life span of seagrass modules, which reflects the life span of their associated meristems, differs greatly among species. Leaf life span ranges from a few days in *Halophila* species to almost a year in *Posidonia oceanica*, and tends to be shorter than that of land plants (Fig. 3.3; Hemminga *et al.*, 1999). The leaves are often shed by the line of contact between the blade and the leaf sheath, when present, which may remain attached to the rhizomes much longer. Roots are also shed, although there is, at present, very limited information on the life span of seagrass roots. Available information suggests that roots are longer lived than leaves, remaining attached to the plants longer than do leaves (Duarte *et al.*, 1998). The life span of the apical meristem of vertical rhizomes ranges from days (e.g. *Halophila ovalis*) to years (Table 3.1), with *Posidonia oceanica* shoots having a maximum life span of several decades (Marbà *et al.*, 1996a). Dead vertical rhizomes, which lack green leaves, can remain attached to the rhizome for long periods of time (Duarte *et al.*, 1994), but may detach at the neck area, the thinner area located at their basal end (i.e. cladoptosis; Bell, 1991). The life span of the apical meristems of horizontal rhizomes is presently

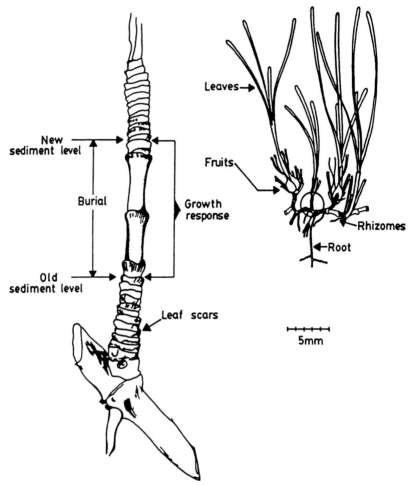

Fig. 3.2. Schematic representation of the scars left on seagrass rhizomes by the loss of seagrass modules, and the changes in length of rhizome internodes in response to external factors, such as seasonality and burial.

unknown, except for species, such as *Zostera*, for which sexual reproduction is a terminal event.

The life span of seagrass modules is scaled to their size, with small species having short leaf life spans and larger species having longer leaf life spans. The existence of an allometric relation between the life span of the modules and seagrass size is to be expected as the efficient use of resources requires that the greater construction costs of larger modules be compensated by a longer functionality.

Table 3.1. *Summary of exponential recruitment and mortality rates and the derived half-life of the shoot population of different seagrass species*

Species	Location	Recruitment (ln units yr^{-1})	Mortality (ln units yr^{-1})	Half-life (years)	Max. age (years)	Reference
Thalassia testudinum	Yucatan Peninsula, Mexico	0.27–1.03	0.6–0.75	1.1–1.4	9.00	Gallegos et al. (1993)
	Florida Bay, USA	0.39–1.51	0.37–2.18	0.3–1.6		Durako (1994)
Thalassodendron cillatatum	Chale lagoon Kenya	0.65	0.71	0.97	5.00	Duarte et al. (1996)
Posidonia oceanica	Spanish Mediterranean	0.02–0.44	0.06–0.53	1.5–11.2	5.2–29.5	Marbà et al. (1996)
Cymodocea rotundata	Bolinao, The Philippines	1.08	0.8	0.78	4.14	Vermaat et al. (1995)
Cymodocea serrulata	Bolinao, The Philippines	3.1	2.86	0.52	1.70	Vermaat et al. (1995)
Enhalus acoroides	Bolinao, The Philippines	0.26	0.23	1.56	9.76	Vermaat et al. (1995)
Halodule uninervis	Balinao, The Philippines	2.06	1.81	0.37	1.26	Vermaat et al. (1995)
Halophila ovalis	Bolinao, The Philippines			0.06	0.22	Vermaat et al. (1995)
Syringodium isoetifolium	Bolinao, The Philippines	2.05	2.46	0.64	1.64	Vermaat et al. (1995)
Thalassia hemprichii	Bolinao, The Philippines	0.77	0.48	1.60	6.07	Vermaat et al. (1995)
Cymodocea nodosa	Ria Formosa, Portugal	0.58–1.48	0.99–3.70	0.18–0.70	2.9–7.6	Cunha (1994)
	Alfacs Bay, Spain	0–3.6	0.1–4.47		7.00	Duarte & Sand-Jensen (1990b)
Halodule wrightii	Yucatan Peninsula, Mexico	4.18	3.54	0.31	3.16	Gallegos et al. (1994)
Syringodium filiforme	Yucatan Peninsula, Mexico	0.77	2.04	0.50	7.66	Gallegos et al. (1994)
Amphibolis antarctica	Perth area, W. Australia	0.61	0.58	1.11	4.30	Marbà & Walker (1999)
Amphibolis griffithii	Perth area, W. Australia	0.34	1.08	1.47	4.58	Marbà & Walker (1999)
Heterozostera tasmanica	Perth area, W. Australia	1.93	2.26	0.93	1.73	Marbà & Walker (1999)
Posidonia angustifolia	Perth area, W. Australia	0.76	1.22	1.01	4.23	Marbà & Walker (1999)
Posidonia australis	Perth area, W. Australia	0.46	0.37	1.79	13.41	Marbà & Walker (1999)
Posidonia sinuosa	Perth area, W. Australia	0.18–0.33	0.45–0.5	2.3–3.0	7.1–10.3	Marbà & Walker (1999)
Thalassodendron pachyrhizum	Perth area, W. Australia	0.5	0.43	2.11	6.41	Marbà & Walker (1999)
Zostera marina	Limfjorden, Denmark	0.11–2.81	0–2.2	31–662		Olesen & Sand-Jensen (1994a)

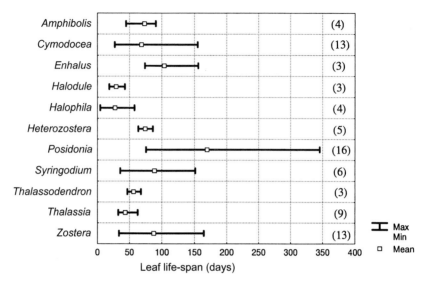

Fig. 3.3. Leaf life span of different seagrass genera. Number of observations in brackets. (Hemminga *et al.*, 1999.)

3.4 Seagrass shoot demography

Although the life span of seagrass clones may extend over millennia (Reusch *et al.*,1999), that of their individual ramets is finite, with shoot life spans ranging from weeks to decades. The mechanisms causing the death of individual shoots, which involves the death of their apical meristems, are poorly known. While the effects of disturbances, such as burial (Marbà & Duarte, 1995; Duarte *et al.*, 1997), acute shading (Neverauskas, 1988; Gordon *et al.*, 1994) and strongly reducing conditions in the sediments (Terrados *et al.*, 1999) on shoot mortality are well established, the causes of the death of seagrass shoots in undisturbed meadows remain obscure. Provided that horizontal meristems, which recruit new shoots to the population, must remain active for seagrass survival (Tomlinson, 1974), the mortality of seagrass shoots is a necessary condition to avoid overcrowding of the population. Indeed, the maximum shoot density of seagrass populations increases with decreasing shoot size (Duarte & Kalff, 1987), as predicted by the self-thinning rule (Westoby, 1986) suggesting that density-dependent regulation is involved. The shoot mortality associated to the self-thinning process may be partially avoided by the close allometric rules that govern the clonal architecture of seagrasses (Chapter 2).

The maintenance of seagrass clones requires a continuous production of new ramets to replace those that die. Small seagrass species, with shoots living for only a few months, must support a much greater shoot production (e.g. Bigely & Harrison, 1986; Vermaat & Verhagen, 1996) than large seagrass species, with long-lived shoots (e.g. Marbà *et al.*, 1996b). Hence, small seagrass species have greater branching rates, faster horizontal rhizome growth, and set consecutive shoots at closer intervals along the growing rhizomes than do large species (Duarte, 1991; Marbà & Duarte, 1998; Chapter 2). In temperate regions, shoot mortality and recruitment are to some extent segregated temporally, with a negative balance between losses and gains leading to shoot population decline in winter. A positive balance between recruitment and mortality, coinciding with the improvement of temperature and irradiance conditions in the spring period, leads to a rapid shoot population increase (e.g. Bigely & Harrison, 1986). Shoot recruitment proceeds until the resources necessary for growth are depleted, which explains the observation of increased shoot density in response to nutrient additions (e.g. Pérez *et al.*, 1991; Agawin *et al.*, 1996). Light may also impose a limit to the density of crowded seagrass meadows (Duarte & Kalff, 1987), as predicted by the self-thinning law (Westoby, 1986). Hence, mortality increases during the summer, probably due to crowding effects, and increases further during autumn and winter in temperate meadows (e.g. Bigely & Harrison, 1986; Harrison, 1993; Olesen & Sand-Jensen, 1994a). The extent of seasonality in shoot density, and seagrass biomass, is much less in temperate species with long-lived shoots, which are capable of a more efficient internal economy of resources (Duarte, 1989; Alcoverro *et al.*, 1995, 1997). Seasonality in shoot density is also lower in the tropics, where fluctuations in light and temperature are less pronounced. Hence, the extent of seasonality in the shoot density of seagrass meadows typically declines from high in high-latitude meadows to low in tropical meadows. Seasonal patterns in population dynamics do not simply result from environmental forcing – there is evidence that they are incorporated into internal rhythms as well. In the case of *Posidonia oceanica*, seasonal growth rhythms were maintained for two years under constant conditions in the laboratory (Ott, 1979).

The age structure of seagrass stands is easily established from the distribution of the number of leaves the extant shoots produced over their life span, calculated as the product between the number of standing leaves plus leaf scars on each shoot and the leaf plastochrone interval (Patriquin, 1973; Duarte *et al.*, 1994). The resulting distributions are characteristically log-normal, with a modal age at < 1 yr and a long, exponentially declining tail as shoot age increases (Fig. 3.4). Age distributions are in fact polymodal, with the different modes representing different annual shoot cohorts (Duarte & Sand-Jensen, 1990b; Duarte *et al.*, 1994; Olesen & Sand-Jensen, 1994a). The exponential decline in the frequency of shoots of increasing age resembles the depletion curves characteristic of most angiosperm populations. A more accurate representation of the life span of seagrass shoots is derived from the distribution of their age-at-death, which can be established from the number of leaf scars on dead shoots (Duarte *et al.*, 1994). The life expectancy of seagrass shoots – the mode of the age-at-death distribution – ranges from < 1 month (*Halophila ovalis*, Vermaat *et al.*, 1995; *Zostera japonica*, Bigely & Harrison, 1986) to 11 years for the species examined, with the longer half-lives corresponding to the larger species (Table 3.1). The distribution of the age-at-death of seagrass shoots presents a better defined exponential decline in the frequency of shoots that attain increasingly long life spans (Duarte *et al.*, 1994). This distribution is indicative of a constant mortality probability, independent of shoot age, which greatly simplifies the examination of seagrass demography compared with the more elaborated approaches applied when mortality is stage specific (cf. Cox & Oakes, 1984). Mortality rates have been reported, however, to be size dependent for *Zostera marina* (Olesen & Sand-Jensen, 1994a), with a higher mortality rate of small shoots. The average exponential mortality rate of seagrass populations ranges, for the species for which it has been established (about half of the flora, Table 3.1), from 0.06 to 3.54 ln units year^{-1}. These rates, equivalent to the loss of 3% to 87% of the shoots present per year, imply that half of the shoot population must be replaced within a time span ranging from 0.06 to 11 years (Table 3.1). Disturbances that lead to the mortality of the apical meristems of horizontal rhizomes can, therefore, rapidly destroy seagrass populations.

In steady-state conditions, shoot mortality must be balanced by an equivalent recruitment of new shoots. The specific shoot recruitment rate can be calculated as the natural logarithm of the fraction of the shoots < 1 yr old in the population (Duarte *et al.*, 1994). Once estimates of shoot recruitment and mortality rates are known for a population (cf. Table 3.1), the trend in population density can be forecast from the net rate of population change, equal to the difference between the recruitment and mortality rates. There is indeed an overall balance between shoot recruitment and mortality rates for the seagrass populations for

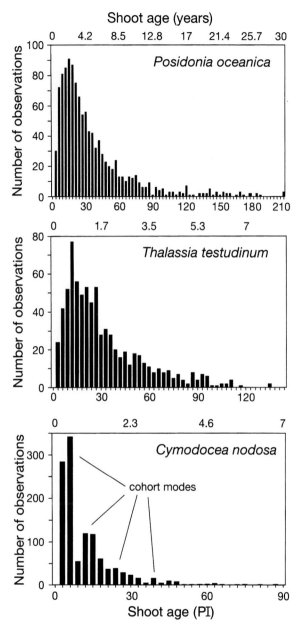

Fig. 3.4. Age structure of seagrass stands, as the frequency distribution of seagrass shoot age (PI, plastochrone intervals). (Duarte *et al.*, 1994.)

which these parameters have been quantified (Duarte & Sand-Jensen, 1990a; Duarte *et al.*, 1994). However, shoot gains and losses are often out of balance for individual populations. An excess recruitment relative to mortality is indicative of expanding populations, which are becoming denser, as has been observed for growing seagrass patches (Duarte & Sand-Jensen, 1990b). In contrast, increased shoot mortality relative to shoot recruitment, as observed in 57% of 29 *Posidonia oceanica* meadows investigated in the Spanish Mediterranean (Marbà *et al.*, 1996a) as well as *Thalassia testudinum* populations undergoing die-off in Florida Bay (Durako, 1994), is indicative of seagrass decline.

A coarse diagnosis on the likely trend of population development is, therefore, possible from simple assessments of seagrass age distribution. The available estimates of shoot mortality and recruitment rates indicate the existence of a negative relationship between these rates and seagrass size (Table 3.1), such that meadows of large seagrass species are characterized by a much slower flux of shoots than those of small, fast-growing seagrass species. Indeed, the production of new shoots is an important component of the net production of small seagrass species, whereas it is negligible for large, long-lived seagrass species (e.g. Marbà *et al.*, 1996b). Lack of accuracy on the age estimates (e.g. Kaldy *et al.*, 1999) as well as uncertainties associated with the extrapolation of inferences derived from specific stations to entire meadows (e.g. Jensen *et al.*, 1996) render these diagnostics crude. Because of the importance of demographic analyses for the prediction of the likely changes in specific seagrass meadows (cf. Chapter 7), the improvement of our capacity to derive reliable estimates of rates of shoot losses and gains should receive due attention in the future.

3.5 Seagrass reproduction

The reproductive biology of seagrass species is relatively well studied, for it has interested naturalists for about two centuries (cf. Pettitt, 1984; McConchie & Knox, 1989). Seagrasses can reproduce both sexually and asexually, through detached or drifting rhizome fragments (e.g. *Zostera marina*), although this process seems to be rare (Ewanchuk & Williams, 1996). The sexual reproduction of seagrasses does not differ from the reproductive modes of their land counterparts except with respect to the frequency of dioecious species and the dispersal processes. Most seagrass genera (9 out of 12) are dioecious, with separate female and male clones, which is relatively rare in land angiosperms (< 10% of the genera). The very high frequency of dioecious plants has been interpreted to be a mechanism to avoid self-fertilization.

3.5.1 *Flowering*

Flowering of seagrasses is often controlled by temperature (McMillan, 1982a; Ramage & Schiel, 1998), and often occurs simultaneously across large spatial scales (Inglis & Lincoln Smith, 1998). Flowering of temperate species typically occurs in late spring, when irradiance improves and water temperature increases, but its timing varies latitudinally (Phillips, *et al.*, 1983). It is also a seasonal event in most tropical species, often associated with the largest spring tides of the year (Pettitt, 1984). The link between seagrass reproduction and tidal amplitude has been known for some time (Pettitt, 1984) and is believed to be linked to the greater transport during spring tides. The Mediterranean species *Posidonia oceanica* is odd amongst temperate seagrass species in that it flowers in the autumn (October; Thelin & Boudouresque, 1985). Remarkably, the timing of reproduction of *P. oceanica* coincides with that of its congeneric species, which are all restricted to the southern hemisphere (Australia). *Enhalus acoroides* is also remarkable amongst seagrasses in that flowering is a more or less continuous process over the year, the individual shoots being able to produce up to 9 flowers each year (Duarte *et al.*, 1998). In contrast, flowering is a rare event for most seagrass species, where typically < 10% of the shoots flower each year (e.g. Johnson & Williams, 1982; Gallegos *et al.*, 1992; Marbà & Walker, 1999; Olesen, 1999), so that many, if not most, shoots do not produce a single flower during their life span, and flowering of some species (e.g. *Cymodocea serrulata*) has only been observed once (Kirkman, 1975). The shoots of many species (e.g. *Cymodocea nodosa, Thalassia hemprichii, Thalassia testudinum, Enhalus acoroides*, and *Posidonia oceanica, Posidonia australis, Thalassodendron pachyrhizum*) have been reported to be able to flower more than once over their life span (Durako & Moffler, 1987; Cox & Tomlinson, 1988; Gallegos, *et al.*, 1992; Duarte *et al.*, 1997; Marbà & Walker, 1999). The possible age dependence of the probability of flowering has been examined for some species, and it appears that shoots of species that live over several years usually do not flower during the first year of life (Gallegos *et al.*, 1992; Duarte *et al.*, 1997). Because of the low probability of flowering, sexual reproduction is a negligible component of the carbon allocation of seagrasses, involving < 10% of the annual production for most species (e.g. Ramage & Schiel, 1998; Olesen,

1999). Again, *Enhalus acoroides* is an exception, for this species devotes more carbon to sexual reproduction than to leaf and rhizome growth (Duarte *et al.*, 1997).

The reproductive effort of seagrasses can be highly variable between years and among populations (e.g. *Posidonia oceanica*; Pergent & Pergent-Martini, 1990), although consistently high flowering among years has been reported for populations of some species (e.g. *Cymodocea nodosa* in the Canary Islands; Reyes *et al.*, 1995). While fruit and seedling production of *Posidonia oceanica* appears to be relatively high in the Ligurian Sea (Balestri *et al.*, 1998b), it is seldom observed along the Spanish Mediterranean coast. Disturbances, such as burial derived from the migration of sand waves or hurricanes, also enhance seagrass flowering (Gallegos *et al.*, 1992; Marbà & Duarte, 1995).

3.5.2 Pollination

All seagrass species, except *Enhalus acoroides*, have hydrophilous pollination, in which pollen grains, usually arranged in pollen strains or search vehicles (McConchie & Knox, 1989), are released in the water column to fertilize the female flower. *Enhalus acoroides* is the only seagrass species with surface pollination. Its pollen grains are released at high tide, and travel on the surface to fertilize the female flower at low tide (Pettitt, 1984). Some seagrass species have remarkably large pollen grains, such as *Enhalus acoroides* and *Amphibolis*, with 3000 μm × 20 μm grains (McConchie & Knox, 1989). The anthers may open underwater releasing the pollen or may break and detach to transport the pollen to the female flower. Seagrass pollen is often negatively buoyant; female flowers in many species have usually short peduncles, and are, therefore, positioned relatively close to the sediment surface. Despite the considerable attention devoted to seagrass pollination, knowledge of the dispersal range of pollen is limited. A study of *Zostera marina* indicated that 95% of the pollen was retained within 15 m from the source (Ruckelshaus, 1996). The pollen and stigma are coated with an adhesive substance which binds strongly upon contact, resembling the two components of epoxy adhesive (McConchie & Knox, 1989).

3.5.3 Seed dispersal

Seagrass seeds are typically negatively buoyant (Pettitt, 1984). The mature seeds of many seagrass species (e.g. *Cymodocea* spp., *Halodule* spp.) are produced at the base of the shoots, and are often positioned at, or just below, the sediment surface. These seeds are, therefore, not likely to disperse far. In contrast, the seeds of some other species are able to disperse because of the presence of structures which confer them buoyancy, such as the buoyant seed coats of *Enhalus acoroides*, or the buoyant seed-containing spathes of some species (e.g. *Zostera* and *Syringodium* species; Curiel *et al.*, 1996). The genera *Amphibolis* and *Thalassodendron* are viviparous; the seedlings develop still attached to the mother plant, from which they acquire resources (Paling & McComb, 1994). Indeed, the seedling is also a dispersal stage for many seagrass species, notably *Thalassia* spp., which can travel considerable distances propelled by currents. The dispersal range of seagrass seeds is a very poorly studied aspect of their reproductive ecology. Although the seeds of many species, including the viviparous species and those whose seeds are set close to or within the sediment, do not disperse far (e.g. Terrados, 1993), robust estimates of dispersal events are only available for *Zostera marina* populations. The available estimates indicate that 95% of the seeds are retained within 30 m from the source (Ruckelshaus, 1996), and 80% of the seeds produced by a population growing in Chesapeake Bay remain within 5 m from the position where they were broadcast (Orth *et al.*, 1994).

Once in the sediment, the seeds of some seagrass species can remain dormant for some time, with a documented dormancy period of 1–2 months for *Zostera marina* (Moore *et al.*, 1993) and 7–9 months for *Cymodocea nodosa* (Cayé & Meinesz, 1985; Reyes *et al.*, 1995), thereby building a rather ephemerous seed bank. Indeed, seed survival for more than one year appears to be a rare situation in seagrasses (e.g. Moore *et al.*, 1993). While high seed production (e.g. > 50 000 seeds m^{-2} yr^{-1} for some stands; Kuo *et al.*, 1993; Fishman & Orth, 1996) and dense seed banks (e.g. > 1000 seeds m^{-2} for *Cymodocea nodosa;* Reyes *et al.*, 1995; and for *Zostera marina;* Olesen & Sand-Jensen, 1994b) have been reported in some meadows, seedling density is always comparatively low (typically about 1% to 10% of the seeds produced; cf. Cayé & Meinesz, 1985; Duarte & Sand-Jensen 1990a; Harrison, 1993; Terrados, 1993; Olesen & Sand-Jensen, 1994b; Orth *et al.*, 1994; Reyes *et al.*, 1995; Fishman & Orth, 1996; Olesen 1999), providing evidence of high seed losses. These losses are due to many factors, including lack of viability, physical damage, export to unsuitable areas, and burial (Keddy & Patriquin, 1978; Harrison, 1993). Moreover, seeds are also lost through predation, as documented by the experimentally estimated 65% loss of

Fig. 3.5. Sequential stages of development of *Cymodocea nodosa* seedlings. (Photograph by J. Terrados.)

Zostera marina seeds to predators (Fishman & Orth, 1996). Conditions conducive to seed germination have been tested many times under laboratory conditions (e.g. Hootsmans *et al.*, 1987; Moore *et al.*, 1993; Balestri *et al.*, 1998a), but the controls on germination in the field are not well understood.

3.5.4 Seedling dynamics

Seedling mortality is very high, with a reported survival probability of $< 2\%$ over the first year for most species (Harrison & Bigely, 1982; Harrison, 1993; Duarte & Sand-Jensen, 1990a; Olesen & Sand-Jensen, 1994b, Olesen, 1999). This represents another important bottle-neck in the life cycle of seagrasses. Seedling survival may be occasionally very high, as the exceptionally high seedling survival rates reported for *Posidonia oceanica* seedlings established on dead rhizome mats (Balestri *et al.*, 1998b). Seedlings invest most of the resources received from the mother plant in developing a root system (Fig. 3.5), needed to acquire their own resources. This may require a few months (Duarte & Sand-Jensen, 1996), over which there is a net loss of carbon due to respiration, which may result in seedling death. Viviparous species circumvent this

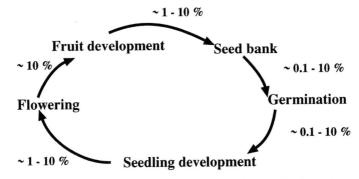

Fig. 3.6. Bottlenecks for the successful sexual reproduction of sea-grasses. Percentages represent indicative values derived from the literature discussed in the chapter.

bottle-neck by transferring resources from the mother plant over the early life of the attached seedling (Paling & McComb, 1994). At a certain stage the seedling falls to the sediment already equipped with the capacity to have a net carbon production. However, they achieve this at the expense of the capacity to disperse at the seed stage.

The preceding discussion on reproduction suggested that successful recruitment from sexual reproduction is a rare event in most seagrass species. Low flowering probability and low survival rates of seeds are major bottlenecks, so that the probability that one particular shoot will develop a successfully established new genet is well below 10^{-5} for most seagrass species (Fig. 3.6). Even when sexual reproduction culminates in the establishment of a new plant, this process is still an ineffective way to propagate, because of the short dispersal distances that most species seem to exhibit. The inefficiency of sexual reproduction highlights the importance of clonal propagation as the main process responsible for the maintenance of seagrass meadows (e.g. Hootsmans *et al.*, 1987; Harrison, 1993; Curiel *et al.*, 1996; Olesen, 1999; Rasheed, 1999; Reusch *et al.*, 1999) and predicts a tendency for low genetic diversity in seagrass meadows. Indeed, a single clone, more than 1 000 years old, of *Zostera marina* has been identified to occupy over half a hectare and comprise more than one million shoots (Reusch *et al.*, 1999).

New developments, based on the application of molecular tools, have confirmed the notion that some seagrass populations are largely clonal and characterized by poor genetic diversity, as demonstrated for *Amphibolis antarctica* (Waycott *et al.*, 1996) and *Posidonia oceanica* (Capiomont *et al.*, 1996; Procaccini *et al.*, 1996; Procaccini & Mazella,

1998). In contrast, genetic diversity has been shown to be important in populations of *Posidonia australis* (Waycott & Sampson, 1997), *Thalassia testudinum* (Schlueter & Guttman, 1998), *Halodule uninervis* (McMillan, 1982b) and *Zostera marina* (Fain *et al.*, 1992; Alberte *et al.*, 1994). Some of the differences outlined may be attributable to the differential resolution of the techniques used, but the information available suggests that horizontal differentiation in the genetic structure of populations is relatively small compared with that observed along depth gradients. The genetic differences between shallow and deep populations seem to be similar in magnitude to those between populations hundreds of kilometres apart (e.g. Procaccini & Mazella, 1998). Genetic variability of *Posidonia oceanica* is remarkably poor, and with low-power techniques it is impossible to falsify the hypothesis that the entire west Mediterranean contains a single clone (Procaccini *et al.*, 1996). More powerful techniques have, however, been able to detect genetic variability within meadows of *Posidonia oceanica*, confirming the overall pattern of a vertical segregation between shallow and deep populations (Procaccini & Mazella, 1998). The horizontal variability of this species has been found to be particularly high along the Alborán–Orán front (Duarte *et al.*, unpubl. results), the oceanographic boundary which represents the true boundary between Atlantic and Mediterranean surface waters, and the biogeographic boundary of *Posidonia oceanica*.

3.6 Seagrass colonization and the seagrass landscape

The coastline that represents the habitat of seagrasses is characterized by a high incidence of natural and human-caused disturbances. Seagrasses can tolerate moderate disturbance through morphological and physiological adaptations, but, if sufficiently strong, disturbance causes seagrass loss. This loss may involve a thinning of the meadow, allowing the recovery through clonal growth from the surrounding rhizome apices, but most often it involves partial or total losses over particular areas of the meadows. In such cases, clonal growth alone is insufficient to allow the prompt recovery of the lost area, and recolonization from seed is necessary (Duarte & Sand-Jensen, 1990a). The bottlenecks for the successful reproduction of seagrasses are many (Fig. 3.6), and the rate of formation of new seagrass patches is much lower than the rate of recruitment of seedlings. In addition to the generally low density of seedlings and their high mortality rate (e.g. Duarte & Sand-Jensen,

Fig. 3.7. A runner (horizontal rhizome extending centrifugally from the patch edge) of *Cymodocea nodosa* on the south-east Spanish coast. (Photograph by J. Terrados.)

1990a; Vidondo *et al.*, 1997; Olesen, 1999), only a few of the surviving seedlings acquire sufficient resources to allocate them to promote clonal growth (Duarte & Sand-Jensen, 1996). The rate of formation of new seagrass patches has been examined for a few seagrass populations, varying from 5×10^{-3} patches ha^{-1} yr^{-1} in a shallow *Zostera marina* population (Olesen & Sand-Jensen, 1994b), to 5×10^{-3} patches m^{-2} yr^{-1} in a *Cymodocea nodosa* population (Duarte & Sand-Jensen, 1990a), and only 3×10^{-4} patches ha^{-1} yr^{-1} in a *Posidonia oceanica* population recovering from bomb blastings (Meinesz & Lefèvre, 1984).

Established patches grow by the horizontal extension of the rhizomes. These extend out of the patches centrifugally (Fig. 3.7), growing faster (because of the production of longer internodes while maintaining the number of internodes each meristem produces annually) when colonizing unvegetated substrate than when they grow within the interior of the meadow. These fast horizontal rhizome growth rates are supported by the translocation of resources from the older shoots along the rhizome (Duarte & Sand-Jensen, 1996). The growth rate of seagrass patches follows a self-accelerating pattern, such that the number of shoots they contain increases exponentially to the 2.3 power of the time elapsed (Vidondo *et al.*, 1997). This exponential increase is similar to that predicted from branching processes when the clonal growth of seagrasses is simulated (Marbà & Duarte, 1998, see section 2.5.1).

It is difficult to provide an estimate of the recovery rate of seagrass species based on their growth capacity. However, estimates of the time scales for seagrass recovery have been produced using simulation techniques. In one such model, the interplay between the horizontal growth and the rate of formation of new patches was combined to provide a first-order estimate of the likely colonization time of different seagrass species. The model results predict recovery times ranging from < 1 year to centuries for seagrasses (Fig. 3.8), with the recovery time being more sensitive to rhizome elongation than to patch formation rates (Duarte, 1995). The predicted recovery times for fast-growing species (*Halophila*, *Syringodium* and *Cymodocea* species) are particularly short, so that even with a moderate rate of new patch formation, a meadow can be established – provided suitable habitat is available – within 1 year (Duarte, 1995). In contrast, the recovery time exceeds a century for *Posidonia oceanica*, forming new patches at the slow rate reported in the Bay of Villefrance (France; Meinesz & Lefèvre, 1984). Hence, *P. oceanica* beds would be in a permanent stage of overall decline if the return time of disturbance is shorter than a century. These simple models assume the rate of patch formation to remain constant, whereas in fact patch formation is expected to increase through time as the growing patches increasingly contribute propagules to the area. Similar models have been found to underestimate the colonization rate of Australian seagrasses (Kendrick *et al.*, 1999), suggesting that, as observed for clonal growth (Marbà & Duarte, 1998), self-accelerating mechanisms may increase the rate of colonization, shortening the time scales for recovery.

The resulting seagrass meadows can be extensive and homogeneous. However, seagrass meadows often display heterogeneity that either reflects that of the environment or results from internal processes derived from seagrass growth. For instance, the meadows of different seagrass species often present circular structures, with low shoot density inside the rings (e.g. Den Hartog, 1971; Fonseca & Bell, 1998). Such circular structures may result from blow-outs, or erosive events related to strong storms (Patriquin, 1975), revealing a link between hydrodynamic activity and landscape attributes of seagrass meadows (Fonseca & Bell, 1998). A threshold of seagrass cover of 50–60% of the substrate, which separates connected seagrass meadows from those where seagrass patches remain discrete (Fonseca & Bell, 1998), has been suggested to separate meadows that remain stable from those that may suffer structural losses during

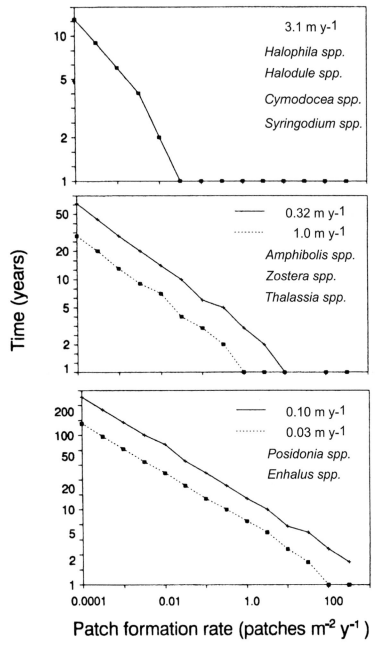

Fig. 3.8. The simulated time to develop dense (95% cover) meadows of seagrass species, as a function of net patch formation rates (*x*-axis) and the different rates of elongation. (Duarte, 1995.)

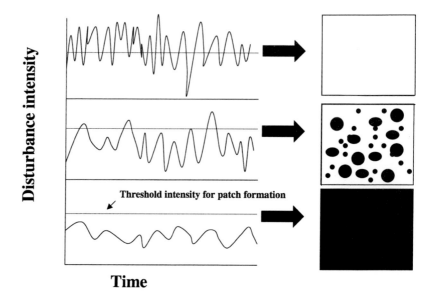

Time

Fig. 3.9. Depiction of the importance of disturbance amplitude and frequency on the development of patchy or continuous meadows, or the maintenance of bare sediments. Perturbations above a certain magnitude cause losses, which, if more frequent than the time needed for meadow formation, maintain a patchy landscape, or, at higher frequencies, result in permanently bare sediments. (Redrawn from Duarte, 1991.)

high-energy periods (Fonseca & Bell, 1998). Circular structures result from the centrifugal growth pattern of some seagrass clones (Den Hartog, 1971), which derives from broad > 30° rhizome branching angles (Marbà & Duarte, 1998), whereby the youngest shoots tend to be located in the periphery of the clone while the old shoots, located in the internal parts of the clone die, leaving the empty space. It has also been suggested that circular seagrass beds are related to topographic features and to the effect of the seagrasses on sediment accumulation (Zieman, 1972).

Some seagrass populations exposed to periodical disturbance (e.g. transit of sand waves) are permanently in a stage of local loss and recovery, resulting in a characteristic patchy landscape resembling a 'leopard skin' (Den Hartog, 1971). Whether the substrate is devoid of seagrass, supports a dynamic patchy landscape, or is a continuous seagrass meadow, depends on the frequency and magnitude of the disturbance relative to the capacity to resist disturbances and the recovery time of the species (Fig. 3.9). Such dynamic equilibrium has been demonstrated

Fig. 3.10. The web of *Cymodocea nodosa* rhizomes becomes exposed as sand waves migrate out of the patch. (Photograph by J. Terrados.)

for *Cymodocea nodosa* growing on a Mediterranean bay swept by sand waves 10–20 cm in height. The progression of the sand waves causes the mortality of the buried patches as the rhizomes become exposed and colonized by drilling organisms following the passage of the sand wave (Fig. 3.10) but the time interval between the passage of consecutive sand waves was sufficient to allow the formation and development of a new patch (Marbà & Duarte, 1995). *Posidonia oceanica* even survives larger (> 50 cm), frequent (> 1 yr^{-1}) sand waves without experiencing mortality (Duarte, unpubl. observations). Hence, a persistent patchy landscape is only possible under combinations of disturbance magnitude and frequency causing only a partial mortality and allowing the partial recovery of the species involved (Fig. 3.11). If the combinations of disturbance magnitude and frequency yield a patchy seagrass landscape can be predicted from knowledge of the patch formation rates and tolerance to disturbance of the species (Marbà & Duarte, 1995).

3.7 Community dynamics of mixed seagrass meadows

The preceding discussion focused on monospecific meadows, which are dominant in temperate waters. However, tropical and subtropical waters

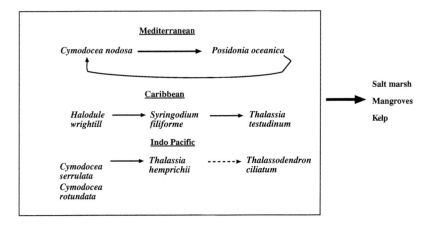

Fig. 3.11. Theoretical successional series proposed for different sea-grass floras. The return line in the Mediterranean series indicates the recolonization of eroded elevated mounds of *Posidonia oceanica*, the dotted line indicates that *Thalassodendron* is believed to represent a terminal stage in the Indo-Pacific series where present, and the external arrow indicates the postulated long-term replacement of the seagrass community by other communities. (After Den Hartog, 1971.)

are often characterized by mixed-species meadows (Chapter 1). Because of the similar architecture and resource requirements of the seagrass flora, coexisting species are expected to show strong interactions, particularly where resources are limiting. The comparison of seagrass communities along ranges of disturbance has led to the postulation of successional series for the different floras (Fig. 3.12), assigning roles as pioneers and climax species to the different species and suggesting the ultimate replacement of the seagrass vegetation by mangroves or salt marshes (Den Hartog, 1971). Examination of the 'climax' and the 'pioneer' species reveals that the former are the large species and the latter the smaller species of the seagrass floras, as expected from the relation between horizontal growth rate and shoot life span, both important components of colonization and maintenance, and seagrass size (Chapter 2).

Perturbation experiments, examining the change in species composition following disturbance, have shown that large, climax species are often sensitive to disturbance (e.g. Duarte *et al.*, 1997). However, some large species can show high resistance to even severe perturbations. For instance, a few *Thalassia testudinum* shoots were able to survive under > 1 m of sand deposited over a Mexican (Yucatan) meadow by Hurricane Gilberto (Marbà *et al.*, 1994), and *Enhalus acoroides* is the South-East

Fig. 3.12. A mixed seagrass meadow (7 species) in the Philippines. (Photograph by C.M. Duarte.)

Asian species that best survives heavy siltation (Terrados *et al.*, 1997). The small, pioneer, species appear to be the first to recolonize the gaps created (Duarte *et al.*, 1997). The species replacement sequences believed to be involved in the succession process during recovery have been tested experimentally in some cases. Disturbance experiments in the Caribbean confirmed that the pioneer species *Halodule wrightii* was replaced by *Syringodium filiforme* and finally by *Thalassia testudinum*, the climax species, during recovery (Williams, 1987, 1990). The postulated successional sequences are, however, seldom realized, for they represent ideal sequences in the absence of disturbance. However, recurrent disturbance appears to be a common feature essential to the maintenance of many seagrass beds. This disturbance may take the form of sedimentary processes, such as sand waves (Marbà & Duarte, 1995), or may involve biotic disturbance, including grazing by macrograzers, such as dugongs, which locally impose a disturbance regime so frequent that *Halophila ovalis*, their preferred food item, and the fastest-growing species in the Indo-Pacific region (Vermaat *et al.*, 1995), is maintained as the dominant species in the community (Preen, 1995; Nakaoka & Aioi, 1999). The successional sequences postulated rely on competition as the factor driving community dynamics, which implicitly requires resource limitation.

Indeed, the long-term experimental addition of nutrients to a mixed Caribbean seagrass bed has been observed to alter the assumed successional sequence for this flora, leading to a dominance of *Halodule wrightii* (Fourqurean *et al.*, 1995). The proposed successional sequence for the Mediterranean Sea proceeds from colonization by *Cymodocea nodosa* to dominance of *Posidonia oceanica*. However, this reef-forming species dies when it develops mounds reaching close to the water surface, which are subsequently colonized by *Cymodocea nodosa* (Den Hartog, 1971), paradoxically believed to represent the terminal stage of the successional sequence. Hence, seagrass successional sequences do not represent unidirectional paths of change, for they may be reverted or short-circuited by disturbance.

In addition to being oversimplified, successional sequences fail to account for the existence of mixed meadows in apparent steady state. Mixed communities often represent steady-state communities in the tropics (Fig. 3.12), and not intermediate stages of a successional sequence. Two types of mechanisms have been described as supporting mixed species meadows: small-scale disturbance and positive interactions among seagrasses. Burrowing thalassinid shrimps are important components of tropical and subtropical seagrass meadows, where their activity disturbs the sediments. These animals have been reported to cause a cumulative disturbance in South-East Asian meadows equal or greater to that of typhoons (Duarte *et al.*, 1997). The mounds and gaps they form (about 20–30 cm in diameter), found in densities of up to $2–3 \ m^{-2}$, are rapidly colonized by small, fast-growing species (Duarte *et al.*, 1997), creating a mosaic of patches at different successional stages. These species act, therefore, as engineer species that maintain mixed meadows through their small-scale disturbance.

While research on seagrasses has emphasized negative interactions, as is the case in plant ecology in general, positive interactions play an important role (Jones *et al.*, 1997). Positive interactions generally involve the activity of species that modify the environment in ways that render it more suitable to support plant life, thereby allowing the growth of other species (Jones *et al.*, 1997). Evidence of positive interactions in seagrass communities has been derived from the comparative analysis of communities growing along gradients of disturbance (siltation) in South-East Asia (Terrados *et al.*, 1997), which showed a positive association between the species richness and the biomass of seagrass communities (Fig. 3.13). The existence of positive interactions was tested experimentally, by conducting series of increasing numbers of species removed following a

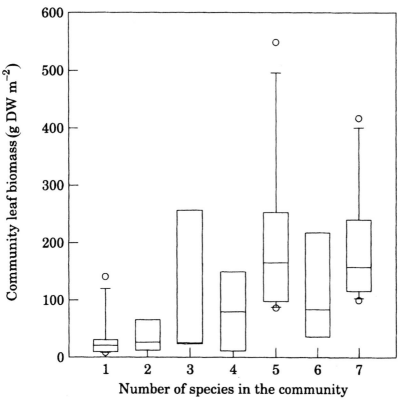

Fig. 3.13. The relationship between species richness and the biomass developed by South-East Asian seagrass meadows. Boxes include the 25% and 75% quartiles of all data, the central line represents the median, bars extend to the 95% confidence limits, and circles represent observations beyond the 95% confidence limits. (Terrados *et al.*, 1997.)

sequence of increasing and decreasing resource use by the species removed in a Philippine mixed-species meadow (Duarte *et al.*, 2000). These experiments suggested that the dominant species (*Thalassia hemprichii*) facilitated the development of denser populations of the accompanying species. These, in turn, sequestered resources that limited the production of the climax species. The nature of the positive interaction was postulated to involve the oxygenation of the sediments by the highly productive climax species, avoiding the accumulation of toxins which were experimentally shown to cause seagrass mortality (Terrados *et al.*, 1999). While the possibility of positive interactions as controlling factors in the community dynamics of seagrasses has only recently been addressed, there are ample opportunities for such interactions to occur, provided that the canopies of large seagrass species will promote sedimentation, thereby enhancing nutrient supply and preventing species growing in their understorey from being washed away. In addition, seagrass seedlings have been shown to survive best within the shelter provided by seagrass meadows than exposed in unvegetated substrate (Balestri *et al.*, 1998b).

3.8 Concluding remarks

Despite the uniform nature of seagrass form, the species differ greatly in the life span of their modules, ranging from days to decades. These differences, linked to the size of the species, impose greatly different population dynamics, with a need for a rapid flux of ramets to maintain meadows of small, short-lived species, to a slower replacement of shoots in meadows of large, slow-growing species. These differences in shoot flux determine the role different species play in the community and the successional series to be observed in the absence of disturbance. Disturbance is, however, a prominent feature of seagrass meadows, allowing the maintenance of complex and diverse communities and spatial patterns, when moderate, but drastically reducing the meadows when intense. Sexual reproduction, which is a sparse process in most seagrasses, becomes the main mechanism to initiate the colonization of bare areas, an extremely slow process in its initial phases, but becomes a self-accelerating process once clonal growth becomes dominant. The time scales involved in the recovery seem to range from a few years for small species up to centuries for long-lived species. Because the dynamics of seagrass populations and communities have important consequences for the associated fauna and biogeochemical processes in the littoral zone, the accurate prediction of these dynamics remains a main challenge for the effective management of these ecosystems.

3.9 References

Agawin, N.S.R., Duarte, C.M. & Fortes, M.D. (1996). Nutrient limitation of Philippine seagrasses (Cape Bolinao, NW Philippines): *in situ* experimental evidence. *Marine Ecology Progress Series*, **138**, 233–43.

Alberte, R.S., Suba, G.K., Procaccini, G., Zimmerman, R.C. & Fain, S.R. (1994). Assessment of genetic diversity of seagrass poupulations using DNA fingerprinting: implications for population stability and management. *Proceedings of the National Academy of Sciences USA*, **91**, 1049–53.

Alcoverro, T., Duarte, C.M. & Romero, J. (1995). Annual growth dynamics of *Posidonia oceanica*: contribution of large-scale versus local factors to seasonality. *Marine Ecology Progress Series*, **120**, 203–10.

Alcoverro, T., Romero, J., Duarte, C.M. & López, N.I. (1997). Spatial and temporal variations in nutrient limitation of seagrass *Posidonia oceanica* growth in the NW Mediterranean. *Marine Ecology Progress Series*, **146**, 155–61.

Balestri, E., Piazzi, L., & Cinelli, F. (1998a). In vitro germination and seedling development of *Posidonia oceanica*. *Aquatic Botany*, **60**, 83–93.

Balestri, E., Piazzi, L. & Cinelli, F. (1998b). Survival and growth of transplanted and natural seedlings of *Posidonia oceanica* (L.) Delile in a damaged coastal area. *Journal of Experimental Marine Biology and Ecology*, **228**, 209–25.

Bell, A.D. (1991). *Plant Form*. New York: Oxford University Press.

Bigely, R.E. & Harrison, P.G. (1986). Shoot demography and morphology of *Zostera japonica* and *Ruppia maritima* from British Columbia, Canada. *Aquatic Botany*, **24**, 69–82.

Brouns, J.J.W.M. (1985). The plastochrone interval method for the study of the productivity of seagrasses; possibilities and limitations. *Aquatic Botany*, **21**: 71–88.

Capiomont, A., Sandmeier, M., Cayé, G. & Meinesz, A. (1996). Enzyme polymorphism in *Posidonia oceanica*, a seagrass endemic to the Mediterranean. *Aquatic Botany*, **54**, 265–77.

Cayé, G. & Meinesz (1985). Observations on the vegetative development, flowering and seeding of *Cymodocea nodosa* (Ucria) Ascherson on the Mediterranean coasts of France. *Aquatic Botany*, **22**, 277–89.

Cebrián, J., Marbà, N. & Duarte, C.M. (1994). Estimating leaf age of the seagrass (*Posidonia oceanica* (L.) Delile) using the plastochrone interval index. *Aquatic Botany*, **49**, 59–65.

Cox, D.R. & Oakes, D. (1984). *Analysis of Survival Data*. New York: Chapman and Hall.

Cox, P.A., & Tomlinson, P.B. (1988). Pollination ecology of seagrass, *Thalassia testudinum*, (Hydrocharitaceae) in St. Croix. *American Journal of Botany*, **75**, 958–65.

Cunha, A. (1994). Aplicaçao das técnicas de reconstruçao ao estudo da dinámica populacional de *Cymodocea nodosa* (Ucria) Ascherson. M.Sc. Thesis. Universidade do Algarve, Faro, Portugal. 98 pp.

Curiel, D., Bellato, A., Rismondo, A. & Marzocchi, M. (1996). Sexual reproduction of *Zostera noltii* Hornemann in the lagoon of Venice (Italy, north Adriatic). *Aquatic Botany*, **52**, 313–18.

Den Hartog, C. (1971). The dynamic aspect in the ecology of sea-grass communities. *Thalassia jugoslavica*, **7**, 101–12.

Duarte, C.M. (1989). Temporal biomass variability and production/biomass relationships of seagrass communities. *Marine Ecology Progress Series*, **51**, 269–76.

Duarte, C.M. (1991). Variance and the description of nature. In *Comparative Ecology of Ecosystems: patterns, mechanisms, and theories*, ed. J.J. Cole, G. Lovett, & S. Findlay, pp. 301–318. Springer-Verlag, New York.

Duarte, C.M. (1995). Submerged aquatic vegetation in relation to different nutrient regimes. *Ophelia*, **41**, 87–112.

Duarte, C.M. & Kalff, J. (1987). Weight-density relationships in submerged macrophytes: the importance of light and plant geometry. *Oecologia*, **72**, 612–17.

Duarte, C.M. & Sand-Jensen, K. (1990a). Seagrass colonization: patch formation and patch growth in *Cymodocea nodosa*. *Marine Ecology Progress Series*, **65**, 193–200.

Duarte, C.M. & Sand-Jensen, K. (1990b). Seagrass colonization: biomass development and shoot demography in *Cymodocea nodosa* patches. *Marine Ecology Progress Series*, **67**, 97–103.

Duarte, C.M. & Sand-Jensen, K. (1996). Nutrient constraints on establishment from seed and on vegetative expansion of the Mediterranean seagrass *Cymodocea nodosa*. *Aquatic Botany*, **54**, 279–86.

Duarte, C.M., Marbà, N, Agawin, N., Cebrián, J., Enríquez, S., Fortes, M.D., Gallegos, M.E., Merino, M., Olesen, B., Sand-Jensen, K., Uri, J. & Vermaat, J. (1994). Reconstruction of seagrass dynamics: age determinations and associated tools for the seagrass ecologist. *Marine Ecology Progress Series*, **107**, 195–209.

Duarte, C.M., Hemminga, M.A. & Marbà, N. (1996). Growth and population dynamics of *Thalassodendron ciliatum* in a Kenyan back-reef lagoon. *Aquatic Botany*, **55**, 1–11.

Duarte, C.M., Uri, J., Agawin, N.S.R., Fortes, M.D., Vermaat, J.E. & Marbà, N. (1997). Flowering frequency of Philippine seagrasses. *Botanica Marina*, **40**, 497–500.

Duarte, C.M., Merino, M., Agawin, N.S.R., Uri, J., Fortes, M.D., Gallegos, M.E., Marbà, N. & Hemminga, M.A. (1998). Root production and belowground seagrass biomass. *Marine Ecology Progress Series*, **171**, 97–108.

Duarte, C.M., Terrados, J., Agawin, N.S.R. & Fortes, M.D. (2000). An experimental test of the occurrence of competition among SE Asian seagrasses. *Marine Ecology Progress Series*, **197**, 231–40.

Durako, M.J. (1994). Seagrass die-off in Florida Bay: changes in shoot demographic characteristics and population dynamics. *Marine Ecology Progress Series*, **110**, 59–66.

Durako, M.J. & Moffler, M.D. (1987). Factors affecting the reproductive ecology of *Thalassia testudinum* (Hydrocharitaceae). *Aquatic Botany*, **27**, 79–95.

Ewanchuk, P.S. & Williams, S.L. (1996). Survival and re-establishment of vegetative fragments of eelgrass (*Zostera marina*). *Canadian Journal of Botany*, **74**, 1584–90.

Fain, S.R., DeTomaso, A. & Alberte, R.S. (1992). Characterization of disjunct populations of *Zostera marina* (eelgrass) from California: genetic differences resolved by restriction-fragment length polymorphisms. *Marine Biology*, **112**, 683–89.

Fishman, J.R. & Orth, R.J. (1996). Effects of predation on *Zostera marina* L. seed abundance. *Journal of Experimental Marine Biology and Ecology*, **198**, 11–26.

Fonseca, M.S., and Bell, S.S. (1998). Influence of physical setting on seagrass landscapes near Beaufort, North Carolina, USA. *Marine Ecology Progress Series*, **171**, 109–21.

Fourqurean, J.W., Powell, G.V.N., Kenworthy, W.J. & Zieman, J.C. (1995). The effects of long-term manipulation of nutrient supply on competition between the seagrasses *Thalassia testudinum* and *Halodule wrightii* in Florida Bay. *Oikos*, **72**, 349–58.

Gallegos, M.E., Marbà N., Merino, M., & Duarte, C.M. (1992). Flowering of

Thalassia testudinum Banks ex König in the Mexican Caribbean: age-dependence and interannual variability. *Aquatic Botany*, **43**, 249–255.

Gallegos, M.E., Merino, M., Marbà., N. & Duarte, C.M. (1993) Biomass and dynamics of *Thalassia testudinum* in the Mexican Caribbean: elucidating rhizome growth. *Marine Ecology Progress Series*, **95**, 185–92.

Gallegos, M.E., Merino, M., Rodriguez, A., Marbà, N. & Duarte, C.M. (1994). Growth patterns and demography of pioneer Caribbean seagrasses *Halodule wrightii* and *Syringodium filiforme*. *Marine Ecology Progress Series*, **109**, 99–104.

Gordon, D.M., Grey, K.A., Chase, S.C. & Simpson, C.J. (1994). Changes to the structure and productivity of a *Posidonia sinuosa* meadow during and after imposed shading. *Aquatic Botany*, **47**, 265–75.

Harrison, P.G. (1993). Variations in demography of *Zostera marina* and *Z. noltii* on an intertidal gradient. *Aquatic Botany*, **45**, 63–77.

Harrison, P.G. & Bigely, R.E. (1982). The recent introduction of the seagrass *Zostera japonica* Aschers. and graeb. to the Pacific coast of North America. *Canadian Journal of Fisheries and Aquatic Sciences*, **39**, 1642–8.

Hemminga, M.A., Marbà, N. & Stapel, J. (1999). Leaf nutrient resorption, leaf life span and the retention of nutrients in seagrass systems. *Aquatic Botany*, **65**, 141–58.

Hootsmans, M.J.M., Vermaat, J.E. & Van Vierssen, W. (1987). Seed-bank development, germination and early seedling survival of two seagrass species from The Netherlands: *Zostera marina* L. and *Zostera noltii* Hornem. *Aquatic Botany*, **28**, 275–85.

Inglis, G.J. & Lincoln Smith, M.P. (1998). Synchronous flowering of estuarine seagrass meadows. *Aquatic Botany*, **60**, 37–48.

Jensen, S.L., Robbins, B.D. & Bell, S.S. (1996). Predicting population decline: seagrass demographics and the reconstructive method. *Marine Ecology Progress Series*, **136**, 267–76.

Jones, C.G., Lawton, J.H. & Shachak, M. (1997). Positive and negative effects of organisms as physical ecosystem engineers. *Ecology*, **78**, 1946–57.

Johnson, E.A. & Williams, S.L. (1982). Sexual reproduction in seagrasses: reports for five Caribbean species with details for *Halodule wrightii* Achers. and *Syringodium filiforme* Kütz. *Caribbean Journal of Science*, **18**, 61–70.

Kaldy, J.E., Fowler, N. & Dunton, K.H. (1999). Critical assessment of *Thalassia testudinum* (turtle grass) aging techniques: implications for demographic inferences. *Marine Ecology Progress Series*, **181**, 279–88.

Keddy, J. & Patriquin, D.G. (1978). An annual form of eelgrass in Nova Scotia. *Aquatic Botany*, **5**, 163–70.

Kendrick, G.A., Eckersley, J. & Walker, D.I. (1999). Landscape-scale changes in seagrass distribution over time: a case study from Success Bank, Western Australia. *Aquatic Botany*, **65**, 293–310.

Kenworthy, W.J. & Schwarzschild, A.C. (1998). Vertical growth and short-shoot demography of *Syringodium filiforme* in outer Florida Bay, USA. *Marine Ecology Progress Series*, **173**, 25–37.

Kirkman, H. (1975). Male floral structure in the marine angiosperm *Cymodocea serrulata* (R. Br.) Ascherson & Magnus (Zannichelliaceae). *Botanical Journal of the Linnean Society*, **70**, 267–8.

Kuo, J., Long, W.L. & Coles R.G. (1993). Occurrence and fruit and seed biology of *Halophila tricostata* Greenway (Hydrocharitaceae). *Australian Journal of Marine and Freshwater Research*, **44**, 43–57.

Lamoreaux, R., Chaney, W.R. & Brown, K.M. (1978). The plastochrone index: a review after two decades of use. *American Journal of Botany*, **65**, 586–93.

Marbà, N. & Duarte, C.M. (1995). Coupling of seagrass (*Cymodocea nodosa*) patch dynamics to subaqueous dune migration. *Journal of Ecology*, **83**, 381–9.

Marbà, N. & Duarte, C.M. (1997). Interannual changes in seagrass (*Posidonia oceanica*) growth and environmental change in the Spanish Mediterranean littoral. *Limnology and Oceanography*, **42**, 800–10.

Marbà, N. & Duarte, C.M. (1998). Rhizome elongation and seagrass clonal growth. *Marine Ecology Progress Series*, **174**, 269–80.

Marbà, N., & Walker, D. I. (1999). Growth, flowering, and population dynamics of temperat Western Australian seagrasses. *Marine Ecology Progress Series*, **184**, 105–18.

Marbà, N., Cebrián, J., Enríquez, S. & Duarte, C.M. (1994a). Migration of large-scale subaqueous bedforms measured with seagrasses (*Cymodocea nodosa*) as tracers. *Limnology and Oceanography*, **39**, 126–33.

Marbà, N., Gallegos, M.E., Merino, M. & Duarte, C.M. (1994b) Vertical growth of *Thalassia testudinum*: seasonal and interannual variability. *Aquatic Botany*, **47**, 1–11.

Marbà, N., Duarte, C.M., Cebrián, J., Enríquez, S., Gallegos, M.E., Olesen, B. & Sand-Jensen, K. (1996a). Growth and population dynamics of *Posidonia oceanica* on the Spanish Mediterranean coast: elucidating seagrass decline. *Marine Ecology Progress Series*, **137**, 203–13.

Marbà, N., Cebrián, J., Enríquez, S. & Duarte, C.M. (1996b). Growth patterns of Western Mediterranean seagrasses: species-specific responses to seasonal forcing. *Marine Ecology Progress Series*, **133**, 203–15.

Mateo, M.A., Romero, J., Pérez, M., Littler, M.M. & Littler, D.S. (1997). Dynamics of millenary organic deposits resulting from the growth of the Mediterranean seagrass Posidonia oceanica. Estuarine, Coastal and Shelf Science, 44, 103–10.

McConchie, C.A. & Knox, R.B. (1989). Pollination and reproductive biology of seagrasses. In *Biology of Seagrasses*, ed. A.W.D. Larkum, A.J. McComb & S.A. Shepherd, pp. 74–111. Amsterdam: Elsevier.

McMillan, C. (1982a). Reproductive physiology of tropical seagrasses. *Aquatic Botany*, **14**, 245–58.

McMillan, C. (1982b). Isozymes in seagrasses. *Aquatic Botany*, **14**, 231–43.

Meinesz, A. & Lefèvre, J.-R. (1984). Régénération d'un herbier de *Posidonia oceanica* quarante années après sa destruction par une bombe dans la rade de Villefranche (Alpes-Maritimes, France). In *International Workshop on Posidonia oceanica Beds*, ed. C.F. Boudouresque, J. de Grissac & J. Olivier, pp. 39–44. G.I.S. Marseille: Posidonie.

Moore, K.A., Orth, R.J. & Nowak, J.F. (1993). Environmental regulation of seed germination in *Zostera marina* L. (eelgrass) in Chesapeake Bay: effects of light, oxygen and sediment burial. *Aquatic Botany*, **45**, 79–91.

Nakaoka, M., & Aioi, K. (1999). Growth of the seagrass *Halophila ovalis* at dugong trails compared to existing within-patch variation in a Thailand intertidal flat. *Marine Ecology Progress Series*, **184**, 97–103.

Neverauskas, V.P. (1988). Response of a *Posidonia oceanica* community to prolonged reduction in light. *Aquatic Botany*, **31**, 261–366.

Olesen, B. (1999). Reproduction in Danish eelgrass (*Zostera marina* L.) stands: size-dependence and biomass partitioning. *Aquatic Botany*, **65**, 209–19.

Olesen, B. & Sand-Jensen, K. (1994a). Demography of shallow eelgrass (*Zostera marina*) populations – shoot dynamics and biomass development. *Journal of Ecology*, **82**, 379–90.

Olesen, B. & Sand-Jensen, K. (1994b). Patch dynamics of eelgrass *Zostera marina*. *Marine Ecology Progress Series*, **106**, 147–56.

Orth, R.J., Luckenbach, M. & Moore, K.A. (1994). Seed dispersal in a marine macrophyte: Implications for colonization and restoration. *Ecology*, **75**, 1927–39.

Ott, J.A. (1979). Persistence of a seasonal growth rhythm in *Posidonia oceanica* (L.) Delile under constant conditions of temperature and illumination. *Marine Biology Letters*, **1**, 99–104.

Paling, E.I. & McComb, A.J. (1994). Nitrogen and phosphorus uptake in seedlings of the seagrass *Amphibolis antarctica* in Western Australia. *Hydrobiologia*, **294**, 1–4.

Patriquin, D.G. (1973). Estimation of growth rate, production and age of the marine angiosperm *Thalassia testudinum* König. *Caribbean Journal of Science*, **13**, 111–23.

Patriquin, D.G. (1975) 'Migration' of blowouts in seagrass beds at Barbados and Carriacou, West Indies, and its ecological and geological implications. *Aquatic Botany*, **1**, 163–89.

Pérez, M., Romero, J., Duarte, C.M. & Sand-Jensen, K. (1991). Phosphorus limitation of *Cymodocea nodosa* growth. *Marine Biology*, **109**, 129–33.

Pergent, G. & Pergent-Martini, C. (1990). Some applications of lepidochronological analysis in the seagrass *Posidonia oceanica*. *Botanica Marina*, **33**, 299–310.

Pettit, J.M. (1984). Aspects of flowering and pollination in marine angiosperms. *Oceanography and Marine Biology Annual Reviews*, **22**, 315–42.

Phillips, R.C., McMillan, C. & Bridges, K.W. (1983). Phenology of eelgrass *Zostera marina* L., along latitudinal gradients in North America. *Aquatic Botany*, **15**, 145–56.

Preen, A. (1995). Impacts of dugong foraging on seagrass habitats: observational and experimental evidence for cultivation grazing. Marine Ecology Progress Series, 124, 201–13.

Procaccini, G. & Mazella, L. (1998). Population genetic structure and gene flow in the seagrass *Posidonia oceanica* assessed using microsatellite analysis. *Marine Ecology Progress Series*, **168**, 133–41.

Procaccini, G., Alberte, R.S. & Mazella, L. (1996). Genetic structure of the seagrass *Posidonia oceanica* in the Western Mediterranean: ecological implications. *Marine Ecology Progress Series*, **140**, 153–60.

Ramage, D.L. & Schiel, D.R. (1998). Reproduction in the seagrass *Zostera novazelandica* on intertidal platforms in southern New Zealand. *Marine Biology*, **130**, 479–89.

Rasheed, M.A. (1999). Recovery of experimentally created gaps within a tropical *Zostera capricornii* (Aschers.) seagrass meadow, Queensland, Australia. *Journal of Experimental Marine Biology and Ecology*, **235**, 183–200.

Reusch, T.B.H., Borström, C., Stam, W.T. & Olsen, J.L. (1999). An ancient eelgrass clone in the Baltic. *Marine Ecology Progress Series*, **183**, 301–4.

Reyes, J., Sansón, M. & Alfonso Carrillo, J. (1995). Distribution and reproductive phenology of the seagrass *Cymodocea nodosa* (Ucria) Ascherson in the Canary Islands. *Aquatic Botany*, **50**, 171–80.

Ruckelshaus, M.H. (1996). Estimation of genetic neighbourhood parameters from pollen and seed dispersal in the marine angiosperm *Zostera marina*, L. *Evolution*, **50**, 856–64.

Schlueter, M.A. & Guttman, S.I. (1998). Gene flow and genetic diversity of turtle grass, *Thalassia testudinum* Banks ex König, in the lower Florida Keys. *Aquatic Botany*, **61**, 147–64.

Short, F.T. & Wyllie-Echeverria, S. (1996) Natural and human-induced disturbance of seagrasses. *Environmental Conservation*, **23**, 17–27.

Terrados, J. (1993). Sexual reproduction and seed banks of *Cymodocea nodosa* (Ucria) Ascherson meadows on the southeast Mediterranean coast of Spain. *Aquatic Botany*, **46**, 293–99.

Terrados, J., Duarte, C.M., Fortes, M.D., Borum, J., Agawin, N.S.R., Bach, S., Thampanya, U., Kamp-Nielsen, L., Kenworthy, W.J., Geertz-Hansen, O. & Vermaat, J. (1997). Changes in community structure and biomass of seagrass communities along gradients of siltation in SE Asia. *Estuarine, Coastal and Shelf Science*, **46**, 757–68.

Terrados, J., Duarte, C.M., Kamp-Nielsen, L., Agawin, N.S.R., Gacia, E., Lacap, D., Fortes, M.D., Borum, J., Lubanski, M. & Greve, T. (1999). Are seagrass growth and survival affected by reducing conditions in the sediment? *Aquatic Botany*, **65**, 175–98.

Thelin, I. & Bourdouresque, C.F. (1985). *Posidonia oceanica* flowering and fruiting: recent data from an international inquiry. *Posidonia Newsletter*, **1**, 5–14.

Tomlinson, P.B. (1974) Vegetative morphology and meristem dependence. The foundation of productivity in seagrasses. *Aquaculture*, **4**, 107–30.

Vermaat, J.E. & Verhagen, F.C.A. (1996). Seasonal variation in the intertidal seagrass *Zostera noltii* Hornem.: coupling demographic and physiological patterns. *Aquatic Botany*, **52**, 259–81.

Vermaat, J.E., Fortes, M.D., Agawin, N., Duarte, C.M., Marbà, N. & Uri, J. (1995). Meadow maintenance, growth, and productivity of a mixed Philippine seagrass bed. *Marine Ecology Progress Series*, **124**, 215–55.

Vidondo, B., Middleboe, A.L., Stefansen, K., Lützen, T., Nielsen, S.L. & Duarte, C.M. (1997). Dynamics of a patchy seagrass (*Cymodocea nodosa*) landscape. Size and age distributions, growth and demography of seagrass patches. *Marine Ecology Progress Series*, **158**, 131–38.

Waycott, M. & Sampson, J.F. (1997). The mating system of a hydrophilous angiosperm. *American Journal of Botany*, **84**, 621–5.

Waycott, M., Walker, D.I. & James, S.H. (1996). Genetic uniformity in *Amphibolis antarctica*, a dioecious seagrass. *Heredity*, **76**, 578–85.

Westoby, M. (1986). The self thinning rule. *Advances in Ecological Research*, **14**, 167–225.

Williams, S.L. (1987). Competition between the seagrasses *Thalassia testudinum* and *Syringodium filiforme* in a Caribbean lagoon. *Marine Ecology Progress Series*, **35**, 91–8.

Williams, S.L. (1990). Experimental studies of Caribbean seagrass bed development. *Ecological Monographs*, **60**, 449–69.

Zieman, J.C., Jr. (1972). Origin of circular beds of Thalassia (Spermatophyta: Hydrocharitaceae) in South Biscayne Nay, Florida, and their relationship to mangrove hammocks. *Bulletin of Marine Science*, **22**, 559–74.

4

Light, carbon and nutrients

4.1 Introduction

As outlined in the first chapter, seagrasses are the only angiosperms that are adapted to a marine submerged existence. Basic requirements for growth are similar for terrestrial angiosperms and seagrasses alike. Life in the marine realm, however, implies exposure to environmental conditions that are considerably different in many respects from those in terrestrial habitats, imposing constraints on the availability of some resources, or calling for specific adaptations to acquire others. In this chapter we will focus on environmental resources imperative for growth in seagrasses, i.e. light, inorganic carbon and nutrients, and on the plant properties relevant to their acquisition and use.

4.2 Light

4.2.1 Availability

Photosynthesis provides plants with chemically fixed energy and with carbon skeletons for the variety of biosynthetic processes associated with plant growth and functioning. The penetration of light through natural waters, however, is at least three orders of magnitude less than through air. Light intensity thus rapidly decreases with water depth, and even in clear ocean water virtually no photosynthetically active radiation (PAR; wavelength 350 or 400 to 700 nm) can penetrate beyond a depth of 200 m. Apart from absorption by pure water, particulate and soluble substances also each contribute to the total attenuation of light in the water column. The intensity of absorption varies with the wavelength; the absorption of pure water, for instance, begins to rise as wavelength

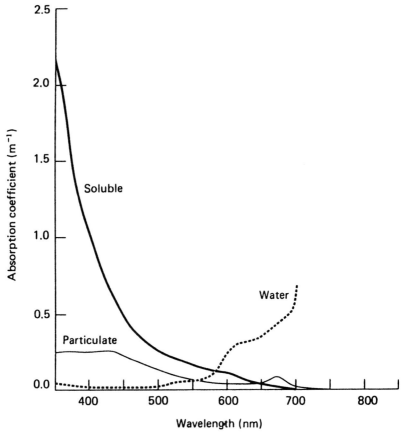

Fig. 4.1. Comparison of the spectral properties of the different fractions in estuarine water (Lake King, Victoria, Australia). (Kirk, 1983.)

increases above 550 nm (Fig. 4.1; Kirk, 1983). Due to the generally higher loads of particles and dissolved organic substances, light transmission in coastal waters is much lower than in clear ocean water. As a consequence the depth limit for penetration of PAR in coastal seas mostly varies from less than a metre to several tens of metres. The distribution of seagrasses thus inevitably is restricted to a narrow depth range. The maximum depth that has been reported for a seagrass is 90 m. This was a *Halophila* species (Taylor, 1928, cited in Den Hartog, 1970), of which the exact species identity is uncertain. The majority of seagrass stands is confined to depths of less than 20 m. There is a general relationship between the colonization depth of seagrasses (Z_c, in m) and

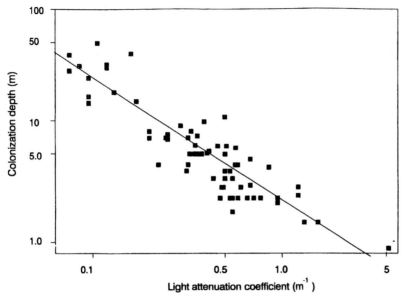

Fig. 4.2. Colonization depth of seagrasses in relation to the water light attenuation coefficient. (Duarte, 1991.)

the light attenuation coefficients (K, in m^{-1}) of the water, derived from a compilation of data (Duarte, 1991):

$$\log Z_c = 0.26 - 1.07 \log K$$

From this equation it can be gathered that colonization depth rapidly decreases as the water becomes more turbid (Fig. 4.2). The data indicate that seagrasses roughly can spread to depths receiving more than 11% surface irradiance.

Unfortunately for coastal management practices, the above-mentioned figure is not directly applicable as an exact definition of the light requirements of individual species, as there exists a marked variability among these. Light requirements may approximate the nearly 11% of surface irradiation emerging as the critical limit derived from the overall data available (Olesen & Sand-Jensen, 1993), but much higher values have also been determined. In *Halodule wrightii* and *Syringodium filiforme*, for instance, minimum light requirements ranged between 24% and 37% of the irradiance measured just beneath the water surface (Kenworthy & Fonseca, 1996). Such differences may partly arise from the different type of methodology applied, but certainly have a species-specific component as well. To complicate matters further, the same species may have different light requirements in different habitats. The reason for this variation is not well understood. Possible explanations are differences in the partitioning between photosynthetic and non-photosynthetic tissues among sites, and that plants may be sensitive not only to total irradiance but also to spectral composition. The wavelength pattern of the light reaching the plants may differ locally due to the variation in the scattering and light absorption characteristics of the particles and dissolved compounds in the water column (Kenworthy & Fonseca, 1996). Furthermore, it has been observed that the light requirements of *Zostera marina* within a single estuary appeared to increase as the variance in light availability increased (Zimmerman *et al.*, 1991). In practice, it is also difficult to determine the relevant value of the attenuation coefficient at a given site. K may be fairly stable in open ocean water, but this is certainly not the case in many coastal environments where water turbidity may fluctuate profoundly as a result of physical and biotic factors. Variable land run-off, phytoplankton blooms, wind and tidal mixing all may combine to produce transient and unpredictable periods of high turbidity alternating with periods of higher water transparency (Cloern, 1996). An example of the fluctuations in water transparency is given in Fig. 4.3, which shows data collected at three estuarine *Halodule wrightii* beds in south Texas (Dunton, 1994). In recent years, it therefore has become increasingly evident that full understanding of the relationship between light availability and seagrass light requirements (and, hence, depth limits) requires site- and species-specific information. Such information is certainly vital for seagrass re-establishment programmes and for assessment of environmental impacts on seagrass beds.

4.2.2 *Photosynthetic characteristics of seagrasses*

For many years seagrass researchers have been intrigued by the issue of how seagrasses cope with the limited light availability underwater. This field got a new impetus in the last decade when it became evident that many of the seagrass declines observed world-wide have been due to deteriorating underwater light conditions. Characteristics of seagrass photosynthesis are primarily determined in laboratory studies as the photosynthetic response of seagrass leaves to increasing light levels. Photosynthetic rates are usually measured as rates of oxygen evolution. Photosynthesis–irradiance (P–I) curves then can be fitted by using one of several available mathematical models. An example of such a curve is

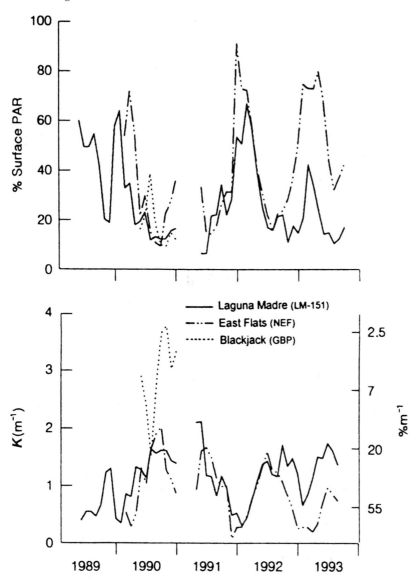

Fig. 4.3. Water transparency at three *Halodule wrightii* beds in south Texas over a 4–year period. Top panel: percentage of surface irradiance reaching the canopy. Bottom panel: attenuation coefficient of the water expressed as K-values (left axis) and as $\% m^{-1}$ (right axis). The seagrass beds were at depths of 1.3 m (LM), 1.2 m (NEF) and 0.6 m (GBP). (Dunton, 1994.)

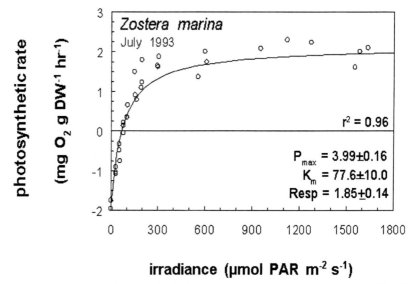

photosynthetic rate (mg O$_2$ g DW^{-1} hr^{-1})

irradiance (μmol PAR m^{-2} s^{-1})

Fig. 4.4. Photosynthesis–irradiance curve of *Zostera marina* (Zandk-reek, south-west Netherlands). The curve was obtained by fitting the data to the model:

$$P = \frac{P_{\mathrm{max}} \cdot I}{(K_{\mathrm{m}} + I)} - R$$

where P is the net photosynthetic rate (mg O$_2$ g DW^{-1} h^{-1}) at irradiance I. P_{max} is the gross maximal photosynthetic rate, R is the respiration rate, and K_{m} is the half saturation constant (μmol PAR m^{-2} s^{-1}). The values of these parameters are given in the figure ± 1 SE. (Vermaat *et al.*, 1997.)

shown in Fig. 4.4. The P–I curves of seagrasses are similar to those of other plants, initially increasing linearly with increasing light, but level-ling off towards a maximum when irradiance becomes saturating. A downward trend after the saturation phase, indicating photoinhibition at higher irradiance levels, has not often been observed. From P–I curves, the maximum photosynthetic rate (P_{max}) can be obtained. P_{max} can either be the gross maximum rate of photosynthesis when the dark respiration term $|\,R\,|$ is added, or the net maximum rate of photo-synthesis without inclusion of the dark respiration term. Other important parameters are the light compensation point (LCP or I_{c}), the irradiance where gross photosynthesis equals respiration and net photosynthesis is zero, and I_{k}, the irradiance indicating the onset of saturation. The initial slope of the relation between P and I is referred to as the photosynthetic efficiency (α).

The large number of studies in which the relation between irradiance and photosynthetic rate has been investigated show that the photosynthetic characteristics vary between species. This is exemplified by a comparative study of nine seagrass species from temperate, Mediterranean and tropical areas showing that P_{max}, R and I_c vary up to fivefold (Vermaat *et al.*, 1997). It is likely that differences between species will have evolved as adaptations to different growth environments. This is, for instance, suggested by the diverging photosynthetic characteristics of the congeneric species *Zostera marina* and *Z. noltii*. These species overlap in their geographic distribution, and can often be found co-occurring or in adjacent zones. A conspicuous difference in their distribution is that *Z. noltii* typically occurs intertidally, whereas *Z. marina* may occur intertidally, but is mostly confined to permanently submerged areas. Vermaat & Verhagen (1996) report observations on co-occurring *Z. noltii* and *Z. marina*, which show that the maximum photosynthetic rate of *Z. noltii* shoots incubated in seawater was higher than that of *Z. marina* (14 mg O_2 g AFDW^{-1} h^{-1} and 6 mg O_2 g AFDW^{-1} h^{-1}, respectively), as was the irradiance levels where light saturation started (236 vs. 78 μE m^{-2} s^{-1}). Both species are also able to maintain photosynthesis if the leaves are exposed to air, as long as the leaves remain covered by a water film. When photosynthetic characteristics of exposed leaves were determined (by measuring gaseous CO_2 exchange), maximum photosynthetic rates and saturation irradiances again were higher in *Z. noltii* (Leuschner & Rees, 1993). Hence, it can be concluded that the photosynthetic apparatus of *Z. noltii* shows a more pronounced adaptation to high light availability, consistent with its general distribution higher in the littoral zone.

If we compare the P_{max} values of different seagrass species published in the comparative study of Vermaat *et al.* (1997) with the values established in terrestrial plants, it appears that they are relatively low. The reported P_{max} values of seagrasses range between 3 and 13 mg O_2 g DW^{-1} h^{-1}, whereas the values for terrestrial C_4 and C_3 plants commonly range between 75 and 175 mg O_2 g DW^{-1} h^{-1}, and between 10 and 75 mg O_2 g DW^{-1} h^{-1}, respectively; P_{max} values of seagrasses are even low compared with the range reported for terrestrial C_3 shade plants (12.5–37.5 mg O_2 g DW^{-1} h^{-1}; data on terrestrial plants derived from a compilation by Larcher, 1995, assuming a photosynthetic quotient of 1.25). The comparatively low values in seagrasses are consistent with the general notion that submerged aquatic plants are characterized by a low photosynthetic capacity. This is probably related to the low

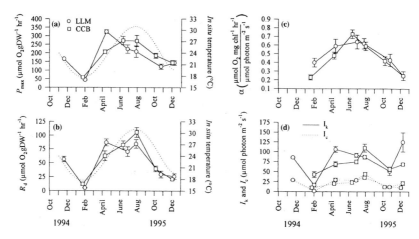

Fig. 4.5. Time course of photosynthetic characteristics of *Thalassia testudinum* at two locations in the Gulf of Mexico (LLM, CCB). (*a*) gross maximum photosynthetic rate, (*b*) dark respiration, (*c*) photosynthetic efficiency, and (*d*) saturation and compensation irradiances (I_k and I_c, respectively). Parameter values derived from laboratory incubations of leaf material (means $\pm$ SE; n = 4 to 6) . Dotted lines indicate approximate *in situ* temperatures during the year. (Herzka & Dunton, 1997.)

chlorophyll content of the leaves per unit of tissue weight compared with terrestrial plants (Agustí *et al.*, 1994), a feature that they share with submerged freshwater angiosperms (Nielsen & Sand-Jensen, 1989). In addition, the supply of inorganic carbon to submerged plants is limited (see section 4.3) restricting photosynthetic activity.

Photosynthetic characteristics are not rigidly fixed within seagrass populations. Seasonal shifts have been observed repeatedly. An example is presented in Fig. 4.5, which shows data on subtropical *Thalassia testudinum* in the Gulf of Mexico (Herzka & Dunton, 1997). P_{max}, R and α peak during the summer followed by declines in autumn and winter. Seasonal changes in photosynthetic characteristics have also been observed in populations of other (temperate and Mediterranean) seagrass species, and point to the general occurrence of seasonal acclimatization of the photosynthetic process to changing environmental conditions (Pirc, 1986; Pérez & Romero, 1992; Terrados & Ros, 1995; Zimmerman *et al.*, 1995; Vermaat & Verhagen, 1996). Among the changes in P–I parameters, higher values of P_{max} in summer than in winter are most frequently observed. In the example of *Thalassia testudinum* given above, the seasonal patterns showed a strong correlation with seasonal shifts in local temperature, suggesting that the seasonal temperature cycle is a factor inducing such changes. Temperature and irradiance of course often show parallel seasonal fluctuations, which makes it difficult to distinguish their effects. It is likely, however, that light conditions can be directly relevant to P–I parameter values, at least in certain seagrass species. This is the obvious explanation for the decrease in net P_{max} that has been observed in *Z. marina* along a depth gradient (Dennison & Alberte, 1986). The coincident finding that reciprocally transplanted eelgrass plants from different depths acquire virtually similar P–I characteristics as not-transplanted control plants corroborates this conclusion. Furthermore, artificial shading of a *Zostera marina* vegetation and experimental lengthening of the photoperiod with underwater lamps led to significant changes in P_{max} (Dennison & Alberte, 1985).

Extrapolation of photosynthetic characteristics determined in laboratory experiments with isolated leaves or leaf segments to the functioning of seagrass vegetation *in situ* should be exercised with some caution. Field-derived photosynthetic parameters may display significantly different values from those determined in the laboratory (Dunton & Tomasko, 1994; Herzka & Dunton, 1997). There may be several reasons for such discrepancies. Respiration rates of leaf segments may be unnaturally high due to wound effects, affecting oxygen production values, and, hence, I_c and I_k values. Laboratory incubation temperature and pressure, factors that can have an important and immediate impact on photosynthetic parameters (Drew, 1979; Beer & Waisel, 1982; Masini & Manning, 1997), are usually not identical to *in situ* conditions. Photosynthetic performance is also dependent on leaf age, and individual leaf segments thus are likely to have characteristics that differ from the characteristics of complete shoots (Alcoverro *et al.*, 1998). The light climate in the canopy of the vegetation, moreover, is much more complex than in the laboratory situation. Apart from the light gradient inherent to light attenuation in the water column, sunlight flickering on the rough water surface, and light absorption and scattering by the moving leaves, continually changes the light conditions within the canopy. Hence, it is hardly possible to imitate the natural light climate in the laboratory.

Chlorophyll *a* fluorescence emitted by green plants reflects various characteristics of the photosynthetic process, and fluorescence measurement techniques have for some time been an important tool enabling plant physiologists to unravel photosynthetic events (Krause & Weis, 1991; see also section 7.6). Recently, technical devices have been developed which allow the *in situ* measurement of chlorophyll fluorescence of submerged plants. The technique has been successfully applied to sea-grasses (Beer *et al.*, 1998; Ralph *et al.*, 1998). An attractive aspect of this technique is that leaf photosynthetic rates under natural conditions can be derived from the measurements, without the need for removal of the leaves and of transfer to laboratory incubation chambers. Application of this technique thus may yield more reliable information on the *in situ* functioning of the photosynthetic system (including rapid responses to changes in the ambient light climate) than the current laboratory O_2 evolution technique.

4.2.3 Light and the carbon balance

The ratio between carbon fixed in photosynthesis and the consumption of organic carbon in respiration is crucial in determining whether the plants show a net positive carbon balance, allowing growth, or not. A positive balance is only achieved at light levels higher than the light compensation point. Laboratory experiments with isolated leaf material has demonstrated that this critical light level for seagrass leaf tissue ranges between 1 and 175 (μmol m^{-2} s^{-1}, but that it usually has a value between 20 and 50 (μmol m^{-2} s^{-1} (Hemminga, 1998). A study of Mediterranean seagrasses and macroalgae has shown that light compensation irradiances tend to increase with increasing tissue thickness (Enríquez *et al.*, 1995), which will be one reason for the observed variability. Apparently, with increasing leaf thickness, the respiratory burden of the leaf tissue increases faster than its ability for carbon fixation. This is in agreement with the general trend observed among a wide range of photosynthetic organisms that the amounts of chlorophyll *a* per unit of tissue weight decrease with leaf thickness (Agustí *et al.*, 1994).

As irradiance levels increase, rates of carbon fixation increasingly outweigh the carbon consumption associated with respiration of the leaf, until the light saturation point is reached. At saturating light levels, the maximum rate of net photosynthetic carbon fixation may exceed respiratory carbon consumption of the leaf tissues some 20 times, but typically this value is around 5 times (Hemminga, 1998). On the scale of the entire plant, however, this surplus on the carbon fixation side of the balance is less favourable due to the additional respiratory burden of the roots and rhizomes. On the basis of oxygen uptake measurements in the laboratory, the respiratory rates of the subterranean tissues per unit dry weight are found to be severalfold (2.4–5) lower than those of the leaf tissues

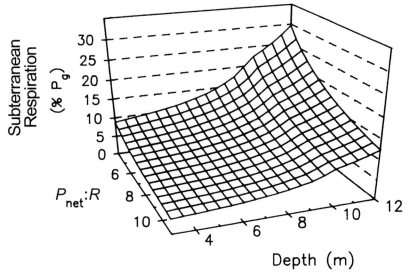

Fig. 4.6. *Zostera marina*. Carbon demand of root and rhizome tissues
(percentage of gross photosynthetic production, i.e. total carbon fixed),
as a function of depth (irradiance) and shoot P_{net}:R. Results pertain to
a subtidal population in Monterey Bay, California, USA. (Kraemer &
Alberte, 1993.)

(Hemminga, 1998). The contribution of the roots and rhizomes to the
total plant biomass, however, can be very high, and may greatly exceed
the leaf biomass. The respiring roots and rhizomes thus may be responsible for a considerable proportion of total plant respiration. Fourqurean
& Zieman (1991), for instance, determined that green photosynthetic
leaves only accounted for 15% of total *Thalassia testudinum* biomass –
roots, rhizomes and belowground parts of the shoots representing the
remaining 85%. The leaves had higher respiration per unit weight, but
this did not outweigh the large share of the belowground structures in
total plant biomass: respiration by the belowground structures comprised
more than half (57%) of total plant respiration. Lower figures have been
calculated for a population of *Zostera marina* in a subtidal meadow
along the coast of California, with a belowground biomass accounting
for a moderate 20–26% of of total plant biomass. In this case, the carbon
respired in the belowground tissues represented less than 15% of the total
carbon fixed at depths < 10 m, where most of the vegetation was found.
At the deepest edges of the meadow, the carbon devoted to respiration of
the roots and rhizomes sharply increased to 30% of the total carbon fixed
(Kraemer & Alberte, 1993; Fig. 4.6).

The respiratory requirements of the belowground tissues, whether they are moderate or more substantial, inevitably will raise the light compensation point for photosynthesis in entire plants above that of the leaves only. Consistent with this supposition, the light compensation point of entire *Halodule wrightii* plants appeared to be 111 μmol m^{-2} s^{-1}, much higher than the 22 μmol m^{-2} s^{-1} determined as the light compensation point for isolated leaf segments (Dunton & Tomasko, 1994). Particularly in low-light environments, where photosynthesis can only proceed at limited rates, it is conceivable that a considerable part of the carbon that is fixed in photosynthesis is counterbalanced by the amount of carbon that is consumed in respiratory processes in the belowground system. Together with additional carbon losses due to excretion and grazing, this severely limits the plants' potential for biomass production.

4.2.4 Light reduction, growth and survival

The effects of light reduction on seagrasses have been investigated in a number of studies in which light intensity was artificially reduced by shading screens. With this type of approach not only the light intensity is reduced, but also the period of light-saturated photosynthesis is diminished or reduced to zero. Both aspects may be important, but for *Zostera marina* in particular it has been argued that it is the period of light-saturated photosynthesis that controls plant responses (Dennison & Alberte, 1985). Whether this is true for other species as well is unknown. Whatever the relative importance of these aspects of the experimental treatment, the responses of seagrasses to light reduction appear to be much the same as in other plant groups. A decrease in leaf production per shoot, which may be accompanied by an increased leaf chlorophyll concentration, is an early response to light reduction (within a month) found in *Zostera marina*, *Z. noltii*, *Posidonia australis* and *Thalassia testudinum* (Dennison & Alberte, 1982; Vermaat *et al.*, 1993; Czerny & Dunton, 1995; Fitzpatrick & Kirkman, 1995). Increased chlorophyll levels have been observed at deep edges of seagrass meadows as well, and are often involved in the acclimatization of plants to low light conditions. If serious light reduction continues over longer periods, shoot density invariably decreases. This phenomenon is reflected in field situations by the decrease in shoot densities towards the depth limit of seagrass meadows (Dawes, 1998).

μThe tolerance to prolonged shading differs among species, with species showing widely different survival capacities under severely reduced light conditions (Gordon *et al.*, 1994; Czerny & Dunton, 1995; Longstaff *et al.*, 1999). Furthermore, the response of a species may vary with the season (Fitzpatrick & Kirkman, 1995). Both phenomena may be related to the variable level of stored carbohydrate reserves in the tissues, notably the rhizomes. In *Thalassia testudinum* growing in the Gulf of Mexico, for instance, non-structural rhizome carbohydrates strongly increase during the summer and drop during winter and early spring, suggesting that these reserves are normally mobilized for the maintenance metabolism of the plants during winter and support of plant growth at the early phase of the growing season (Lee & Dunton, 1997). A similar pattern with a build-up of non-structural carbohydrates during the summer and a decline of these reserves to seasonal lows during the winter has been observed in *Zostera marina* in the San Francisco Bay area (Zimmerman *et al.*, 1995). The same species growing near Chesapeake Bay, however, accumulates reserves in a bimodal pattern during spring and autumn, parallel to the spring and autumn growth peaks in this region (Burke *et al.*, 1996). Consequently, the capacity of the plants to rely on their reserves when carbon compounds are required for maintenance metabolism or growth will vary seasonally. The plants are presumably more vulnerable, and their persistence more at risk, by shading during periods when carbohydrate reserves are at low levels.

4.2.5 *Light reduction and the oxygen supply to roots and rhizomes*

Apart from differences in the amounts of carbohydrate reserves, the development of hypoxic or anoxic conditions in the roots and rhizomes could also play a role in the tolerance of seagrasses to light reduction. An inmediate consequence of light reduction is a reduced production of photosynthetic oxygen. This is of direct relevance to the functioning of the subterranean tissues. Whereas in terrestrial and emergent wetland species the photosynthetic process appears to be only of minor importance in providing oxygen to the belowground tissues, the opposite is the case in seagrasses. The sediment of seagrass beds is usually anoxic, and the oxygen supply of the roots and rhizomes therefore is dependent on an oxygen flux from the shoots to the belowground organs. The oxygen dissolved in the water column would seem to be of limited value in this respect, as the very low diffusivity of gases in water restricts the entry of oxygen into the shoots. Hence, the oxygen available to the roots and rhizomes is largely derived from photosynthesis. This oxygen is conveyed to the roots and rhizomes via the system of internal gas spaces (lacunae) that extend virtually from the leaf to the root apices, the major mechanism for movement through the lacunal airspaces probably being gaseous diffusion (Larkum *et al.*, 1989). The presence of such a system of internal airspaces is typical for plants tolerant of anoxic sediments, and may be more or less developed depending on the reducing conditions of the sediment (Penhale & Wetzel, 1983).

Not all of the oxygen is consumed in the support of aerobic respiration of the belowground tissues: some of it may leak out of the root/rhizome complex. These oxygen losses can be measured relatively easily in laboratory experiments, and give a clear demonstration of the photosynthesis-derived oxygen flow to the belowground organs in seagrasses. Connell *et al.* (1999), for example, measured the effect of shoot illumination on the oxygen release from roots of *Halophila ovalis*, using a movable O_2 electrode that fitted as a sleeve around the root. Oxygen release from the roots rose sharply after illumination of the shoots; however, when the leaves had been removed, no oxygen loss from the roots could be detected during illumination. With the moving electrode it was possible, moreover, to demonstrate that oxygen loss did not occur over the entire root length, but was confined to the root tip. This restriction of the zone of oxygen loss may be important for the plant, as it enhances the diffusive oxygen flux to the root apices, and thus allows the roots to extend further into anoxic sediments.

As light conditions deteriorate, photosynthetic rates decrease, and the oxygen flow to the roots and rhizomes consequently diminishes. If hypoxia or anoxia of the belowground tissues ensues as a result of this reduced flow, it may negatively affect the plant. Anaerobiosis of belowground tissues has been studied into some depth in *Zostera marina*. Incubation of roots under anoxic conditions with ^{14}C-labelled sucrose showed that fermentation of this sugar predominantly leads to the formation of CO_2, lactate and ethanol (Smith *et al.*, 1988). Ethanol was promptly released from the roots into the incubation medium. By this mechanism accumulation of this potentially phytotoxic compound in the tissues is avoided, but at the same time it implies a drain of carbon. In addition, anaerobiosis may affect the utilization of carbohydrate reserves stored by the plants. Further research on eelgrass showed that translocation of sucrose from shoots to roots is inhibited when the roots become anoxic, whereas starch probably is not mobilized in anoxic roots (Zimmerman & Alberte, 1996). This implies that the roots have to rely on their own sucrose pools for ATP generation during anoxic episodes. Exhaustion of sucrose pools following anoxia thus could lead to impairment of root functioning and, ultimately, to plant death, even though starch reserves in the plant tissues are still high.

Whether anaerobiosis actually occurs in seagrass roots and rhizomes during reduced light conditions or in complete darkness is still a matter of speculation. In *Cymodocea rotundata*, laboratory experiments show that an oxygen supply to the roots is maintained even in darkness (Pedersen *et al.*, 1998). In this species, in the absence of photosynthetic activity, oxygen diffusion from the water surrounding the leaves into the lacunal system apparently is sufficient as an oxygen-providing mechanism.

4.2.6 *Light reduction and changes in the shoot to root/rhizome ratio*

When light availability is low, the plants would profit from a reduced formation of belowground biomass in favour of the formation of leaves. This would reduce the maintenance costs of the non-photosynthetic tissues, and at the same time would support carbon fixation through formation of photosynthetic biomass. Such a response is predicted by the concept of 'functional equilibrium' between shoots and roots as originally formulated for terrestrial plants (Brouwer, 1962, 1963), which states that root growth is limited by the rate of supply of carbon from the leaves, whereas the growth of leaves is limited by the supply of water or nutrients by the roots. Evidence that this concept may also apply to seagrasses is given by Olesen & Sand-Jensen (1993). These authors measured growth of leaves and roots/rhizomes of field-collected *Zostera marina* at different irradiance levels. In each of the three seasons that the experiment was carried out, the light compensation point for growth of the leaves was considerably lower than that of the roots and rhizomes (Fig. 4.7). At low irradiance levels leaf growth is maintained whereas at the same time the roots/rhizomes even lose weight, probably due not only to respiration, but also to re-allocation of compounds from these below-ground organs to the leaves. Only at higher irradiance levels did both leaves and roots/rhizomes gain weight. Such growth responses to ambient light, which have also been observed in freshwater macrophytes, can be expected to occur more widely in seagrasses. This needs further verification, however.

4.2.7 *Light and the seasonality of seagrass biomass*

Seagrass meadows exhibit considerable fluctuations in biomass. Most of this variability reflects responses to seasonally changing environmental

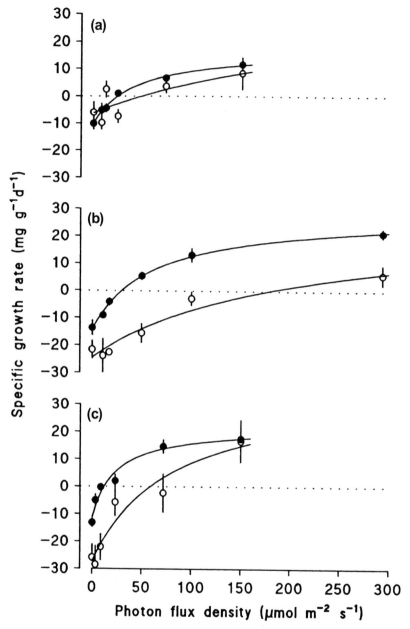

Fig. 4.7. *Zostera marina*. Specific growth rates of leaves (solid symbols) and roots plus rhizomes (open symbols) as a function of irradiance for plants collected in (*a*) March, (*b*) August, and (*c*) October. Means ±95% confidence limits; *n* = 5 to 6). (Olesen & Sand-Jensen, 1993.)

conditions (Duarte, 1989). Light is crucially involved in the pronounced seasonal fluctuations in biomass (and productivity) observed in temperate seagrasses: these are thought to be primarily controlled by seasonal shifts in irradiance and temperature, with individual species differing in their relative sensitivity to these two climatic factors (Marbà *et al.*, 1996; Laugier *et al.*, 1999). Thus, the typical seasonal biomass pattern of temperate seagrasses follows the seasonal changes in solar energy, with a strong increase in spring, a peak value during the summer months, and a subsequent decline in autumn (e.g. Orth & Moore, 1986; Van Lent & Verschuure, 1994; Vermaat & Verhagen, 1996). The more stable light and temperature conditions in tropical regions are reflected in a generally more uniform biomass throughout the year (Hillman *et al.*, 1989; Duarte, 1989). However, considerable seasonal biomass fluctuations have been also observed in the tropics. Moderate changes in daily light period and temperatures, partially explain such observations in some cases (Mellors *et al.*, 1993; Lanyon & Marsh, 1995). In the Indo-Pacific region, seasonal exposure of reef flat meadows during spring low tides that occur during daylight, resulting in the desiccation and burning of the leaves, has been pinpointed as the cause of conspicuous biomass declines (Erftemeijer & Herman, 1994; Stapel *et al.*, 1997).

4.3 Inorganic carbon

4.3.1 Acquisition

Equally essential to photosynthesis as light is the supply of carbon. Carbon constitutes approximately 30–40% of tissue dry weight, being incorporated in the numerous organic compounds that are essential for the physical structure and the metabolic functioning of the plants. An adequate supply of carbon dioxide, the principal form in which carbon is absorbed for photosynthetic fixation in terrestrial plants, however, is less obvious in the marine than in the terrestrial environment. In the first place, this is because carbon dioxide concentrations in seawater are relatively low. Whereas in freshwater the ratio of the concentration of CO_2 in the atmosphere to the concentration in the water (the partition coefficient) is about 1 between 10 °C and 20 °C, the CO_2 solubility in seawater is 10% to 15% lower, due to the effect of the dissolved salts. Seawater at a temperature of 20 °C thus contains only about 12 µM free dissolved CO_2. More important is the fact that diffusion rates for CO_2 in seawater are some 10 000 times slower than in air. This low diffusion rate is an unavoidable restriction of CO_2 supply: even when leaves are submersed in flowing water, there is a stagnant layer immediately adjacent to the leaf surface, the so-called unstirred or diffusion boundary layer. In this layer transfer of CO_2 (or other solutes) occurs only through molecular diffusion. The thickness of the diffusion boundary layer on seagrass leaves varies with the changes of hydrodynamic conditions and the roughness of the leaf surface. In well-stirred solutions the layer may be only 50 μm thick, whereas under natural conditions with medium to low water motion it may be as much as 1000 μm (Larkum *et al.*, 1989). Also the presence of epiphytes, hairs, silt particles, etc., on the leaf surface affects the diffusion boundary layer. A continuous thick epiphyte cover coincides with an expansion of the boundary layer (Sand-Jensen *et al.*, 1985), but protuberances that are more scattered over the leaf surface may contribute to its partial breakdown by promoting the formation of microscale eddies (Koch, 1994).

The obvious physical restrictions of the CO_2 supply to submerged angiosperms has led seagrass researchers to focus on an alternative source of carbon that is particularly abundant in seawater, bicarbonate. In water CO_2 not only dissolves, but it also reacts with it:

$$CO_2 + H_2O \leftrightarrow \underset{\text{carbonic acid}}{H_2CO_3} \leftrightarrow H^+ + \underset{\text{bicarbonate}}{HCO_3^-} \leftrightarrow \underset{\text{carbonate}}{CO_3^{2-}} + 2H^+$$

The equilibrium concentrations of each of the different components of this carbonic acid system in open seawater are widely different (Stumm & Morgan, 1981). By far the major component in seawater is HCO_3^-, which (in seawater with a pH of 8.2, a salinity of 35‰ and a temperature of 15 °C) accounts for about 90% of the carbon, whereas dissolved free CO_2 accounts only for about 0.6%. Around 9% is present as CO_3^{2-} ions, and a negligible amount as undissociated H_2CO_3 (Dring, 1982). The diffusion coefficients of free dissolved CO_2 and HCO_3^- are of the same order of magnitude, implying that, when both components would be depleted at the epidermal leaf surface due to utilization by the cells, the flux of HCO_3^- diffusing through the diffusion boundary layer would be much larger than that of free dissolved CO_2.

To date some 15 species from 9 out of the 12 extant seagrass genera have been investigated with respect to their capacity to utilize HCO_3^- as a source of carbon. Most of these studies used an experimental set-up in which the photosynthetic rates of leaves were measured at different pH levels. Between pH 6 and 9 the proportion of inorganic carbon that is present in seawater as free dissolved CO_2 rapidly declines, and is virtually zero at pH 9. At this high pH, virtually all inorganic carbon is in the form of HCO_3^- or CO_3^{2-}. Higher photosynthetic rates than could be expected on the basis of the diminishing CO_2 concentrations at increasing pH levels, and persistent photosynthetic activity even at pH levels close to 9 have generally been taken as evidence for HCO_3^- utilization. More recently, specific inhibitors of photosynthetic inorganic carbon acquisition have entered the scene. The use of these inhibitors may yield indications of HCO_3^- use and of the mechanism of HCO_3^- utilization at the same time (James & Larkum, 1996; Beer & Rehnberg, 1997; Björk *et al.*, 1997; Invers *et al.*, 1999). The accumulated evidence now available gives little reason to doubt the capacity of the majority of seagrasses to utilize HCO_3^- as an inorganic carbon source next to free dissolved CO_2. This capacity has now been shown in representatives of the genera *Cymodocea, Enhalus, Halodule, Halophila, Posidonia, Syringodium* and *Zostera*. The use of HCO_3^- in representatives of two other genera, *Thalassia* and *Thalassodendron*, is less straightforward. Use of HCO_3^- has been demonstrated in *Thalassia testudinum* (Durako, 1993), but the congeneric species *Thalassia hemprichii* may lack this capacity (Abel, 1984; Björk *et al.*, 1997), just as *Thalassodendron ciliatum* does (Björk *et al.*, 1997).

Whereas an uncharged molecule as CO_2 may permeate the cells by passive diffusion (this is the mechanism that supposedly leads to photosynthetic uptake of CO_2 in seagrasses), this is much more unlikely for the charged HCO_3^- molecule. The negative membrane potential of the plasmalemma would require that for substantial diffusion of HCO_3^- to occur into the cell very strong concentration gradients must exist over the cell membrane; this supply mechanism for inorganic carbon to the cell interior is therefore probably unimportant (James & Larkum, 1996). Supply of CO_2 to the cells following uncatalyzed conversion of HCO_3^- to CO_2 is also probably not relevant, as the rates of this process are very slow. Several studies now strongly suggest that this conversion nonetheless is an important element in the acquisition of inorganic carbon by seagrasses, being catalytically mediated by extracellular/surface bound carbonic anhydrase. The CO_2 that is formed as a result of the dehydration subsequently enters the cell by diffusion. The involvement of carbonic anhydrase emerges from experiments with acetazolamide, a membrane-impermeable inhibitor of this enzyme. Addition of acetazolamide to the incubation media led to a considerable reduction – sometimes more than 50% – of photosynthetic rates in several seagrasses, indicating that extracellular dehydration of HCO_3^- is an important element in their acquisition of inorganic carbon (James & Larkum, 1996; Beer & Rehnberg, 1997; Björk *et al.*, 1997; Hellblom & Björk, 1999; Invers *et al.*, 1999). The exact location of carbonic anhydrase is not known yet, but is probably either bound to the cell wall or to the plasmalemma (James & Larkum, 1996). Observations of photosynthesis in *Zostera marina* in the presence of membrane-associated ATP-ase inhibitors have also yielded some evidence that energy-requiring direct uptake of HCO_3^- may occur in seagrasses (Beer & Rehnberg, 1997). The effectivity of CO_2 acquisition via the extracellular carbonic anhydrase mediated HCO_3^- dehydration is considered less than that of direct HCO_3^- uptake as the concentrations of CO_2 and HCO_3^- at equilibrium are not affected by the enzyme, and are sensitive to even slight pH changes. Particularly in shallow water meadows, pH in the seawater may fluctuate as a result of metabolic processes in the system. Invers *et al.* (1997) measured a daily increase of pH 8.1 to 8.6 in the canopy of a shallow *Cymodocea nodosa* meadow over the course of the day. Such changes will shift the equilibrium concentrations of CO_2 and HCO_3^- in favour of HCO_3^-. As a consequence, the diffusion gradient of CO_2 across the plasma membrane is likely to be lower. The resulting lower inward flux potentially may limit carbon fixation in the cells.

4.3.2 *Supply limitation*

Although the catalyzed dehydration of HCO_3^- due to the presence of extracellular carbonic anhydrase (and, in addition, perhaps transport mechanisms as yet insufficiently known) enlarge the influx of inorganic carbon, carbon limitation of photosynthesis may occur in seagrasses. This can be derived from a number of observations that show that photosynthetic rates increase considerably when the inorganic carbon concentration in the incubation medium is increased above the level of that of natural seawater (Beer & Waisel, 1979; Durako, 1993; Beer & Koch, 1996; Björk *et al.*, 1997; Zimmerman *et al.*, 1997). In contrast, photosynthesis of marine macroalgae is often carbon-saturated (or nearly saturated) in natural seawater, as is exemplified in Fig. 4.8. Beer & Koch (1996) have pointed out that during the Cretaceous when seagrasses first appear in the fossil record, environmental conditions for an adequate supply of inorganic carbon were more favourable for seagrasses. Atmospheric CO_2 levels were higher, and seawater pH was lower. Assuming present-day carbon utilization characteristics, levels of inorganic CO_2 and HCO_3^- may have been saturating when seagrasses

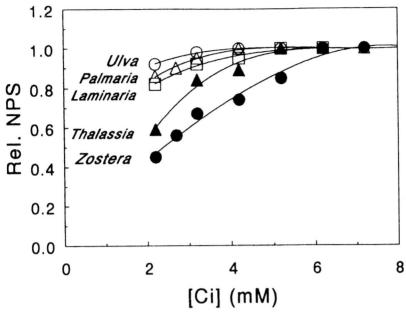

Fig. 4.8. Net photosynthetic rates (NPS) of 2 seagrasses (solid symbols) and 3 marine macroalgae (open symbols) in natural seawater (2.2 mM inorganic carbon, Ci) following additions of Ci. Rates are given relative to the maximal mean rate for each species; the coefficient of variance was less than 10%. (Beer & Koch, 1996.)

started to inhabit the seas, conducive to a high productivity. With the added advantage of roots that allow the utilization of sediment nutrients, this may have facilitated the penetration and colonization of the marine shallow benthic environment, until then monopolized by macroalgae.

Evidence for carbon limitation of photosynthetic carbon fixation under natural present-day conditions is also provided by the considerable body of field data on seagrass stable carbon isotope signatures. Plants generally have a $^{13}C/^{12}C$ ratio in their tissues which differs from the ratio in their inorganic carbon sources, i.e. atmospheric or dissolved carbon dioxide, or bicarbonate. Usually, the $^{13}C/^{12}C$ ratio in plant tissues is lower due to discrimination against the heavy isotope occurring in the sequence of steps leading to the formation of organic plant compounds. The $^{13}C/^{12}C$ ratio of organic C in plants is expressed as the $\delta^{13}C$ value, defined as:

$$\delta^{13}C = \frac{\left(^{13}C/^{12}C\right)_{sample}}{\left(^{13}C/^{12}C\right)_{standard}} - 1$$

where $(^{13}C/^{12}C)_{sample}$ is the isotopic ratio of the sample and $(^{13}C/^{12}C)_{standard}$ is the isotopic ratio of a standard reference material, usually $CaCO_3$ of belemnites from the Cretaceous Pee Dee formation. In the following, $\delta^{13}C$ values are given in relation to this standard.

Marine plants generally have the C_3 type of photosynthetic carbon fixation. This is also true for seagrasses, with *Cymodocea nodosa* possibly being the only exception among the species investigated (Beer, 1989). Plants which employ the C_3 pathway use ribulose 1,5-bisphosphate carboxylase/oxygenase (abbreviated to Rubisco) as the primary carboxylating enzyme. In the chloroplasts, carbon dioxide is bonded covalently to ribulose 1,5-bisphosphate forming a six-carbon compound that immediately splits to form two molecules of 3-phosphoglycerate. The catalytic activity of Rubisco potentially coincides with considerable discrimination against carbon dioxide containing the heavy isotope ^{13}C, with the consequence that the organic compounds formed during the photosynthetic process contain a lesser proportion of ^{13}C than the source carbon dioxide. Whether Rubisco is indeed capable of exerting its pronounced ^{13}C discriminating effect depends on the supply of carbon dioxide. When virtually *all* carbon dioxide molecules that are supplied to Rubisco are used in the carboxylation reaction, no isotopic fractionation between source and product is possible. In terrestrial plants, this situation may be encountered when the diffusion supply of CO_2 through the stomatal pores is low compared with the capacity for carbon fixation. As a result, isotopic discrimination then is mainly determined by the differential diffusivity of $^{13}CO_2$ and $^{12}CO_2$ across the stomatal pathway (Farquhar *et al.*, 1989). Whereas Rubisco may exert an isotopic discrimination of 29‰ to 30‰ relative to the source CO_2, the extent of discrimination associated with diffusion of gaseous CO_2 through the stomatal pore is less pronounced, being 4.4‰. Terrestrial plants with the C_3 type of carbon fixation typically have a $\delta^{13}C$ value of around $-28‰$. As atmospheric CO_2 has a $\delta^{13}C$ value of $-7.8‰$, this implies a 20‰ shift relative to the source carbon, due to the combined fractionation effects of diffusion and carboxylation. For a quantitative partitioning of the contribution of both processes, we refer to the review paper by Farquhar *et al.* (1989). Suffice to say that the characteristically depleted $\delta^{13}C$ values of terrestrial C_3 plants reflect much, but not all, of the discrimination potential of Rubisco.

The natural $\delta^{13}C$ values of seagrasses range between $-3‰$ and $-23.8‰$, but values between $-10.0‰$ and $-11.0‰$ are found most

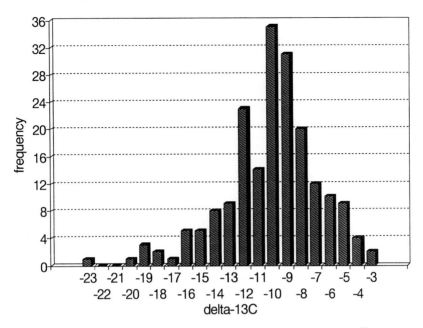

Fig. 4.9. Frequency distribution of published seagrass $\delta^{13}C$ values. (Hemminga & Mateo, 1996.)

frequently (Hemminga & Mateo, 1996, Fig. 4.9). These $\delta^{13}C$ values indicate that seagrasses, being C_3 plants, are remarkably rich in the heavy isotope ^{13}C. Interpretation of the $\delta^{13}C$ value of marine plants is complicated by the fact that potentially more sources of inorganic carbon are available: next to dissolved CO_2, HCO_3^- is an important alternative. Dissolution of atmospheric CO_2 leads to a slight drop in $\delta^{13}C$, giving CO_2 in seawater a $\delta^{13}C$ of about $-9‰$. In contrast, hydration of CO_2 coincides with a considerable ^{13}C *enrichment* relative to the source carbon, giving HCO_3^- in the seas a value of approximately $0‰$. Use of this 'heavy' HCO_3^- in theory could increase the $\delta^{13}C$ values of seagrasses relative to those of terrestrial C_3 plants using 'light' atmospheric CO_2. However, as discussed previously, the present evidence indicates that use of HCO_3^- by seagrasses primarily occurs via the extracellular enzymatic conversion of HCO_3^- to CO_2, followed by diffusive entry of CO_2 into the cells. This carbonic-anhydrase-catalyzed conversion of HCO_3^- to CO_2 yields dissolved CO_2 with a $\delta^{13}C$ which is again close to that of the bulk CO_2 in the water, as the kinetic fractionation of the enzyme-catalyzed conversion of bicarbonate to carbon dioxide is closely similar to the equilibrium fractionation associated with the bicarbonate:carbon dioxide interconversion (Paneth & O'Leary, 1985). Thus, the (indirect) use of HCO_3^- does not seem to offer an adequate explanation for the generally high $\delta^{13}C$ values of seagrasses. Maberly *et al.* (1992), in a discussion of $\delta^{13}C$ values in marine plants, argue that if the source carbon available to marine plants is restricted to dissolved CO_2 with a $\delta^{13}C$ of $-10‰$, complete limitation of photosynthetic rates by Rubisco would yield biomass with a $\delta^{13}C$ of about $-39‰$ ($-10‰$ plus the contribution of Rubisco, $-29‰$). In contrast, complete diffusion limitation would produce biomass with a $\delta^{13}C$ of about $-11‰$ ($-10‰$ plus the contribution of isotope discrimination associated with diffusion of dissolved CO_2 in water, $-0.7‰$). If seagrasses similarly rely on dissolved CO_2, this reasoning suggests that, with the typical $\delta^{13}C$ value of seagrasses being -10 to $-11‰$, carbon fixation is generally limited by the restricted diffusion supply of CO_2 to Rubisco. Consistent with this idea is the observation that $\delta^{13}C$ values of seagrasses become less negative (indicating less discrimination against ^{13}C), when productivity (carbon demand) increases (Grice *et al.*, 1996).

It should be added that, although a limited CO_2 supply may explain much of the enriched $\delta^{13}C$ values of seagrasses, it will not be the complete story as even considerably higher values than the $-10‰$ to $-11‰$ have been reported. These cannot yet be explained adequately.

4.4 Nutrients

4.4.1 Requirements

Some 15 nutrient elements are required by all plants. N, P, S, K, Ca and Mg are required in considerable amounts whereas others such as Fe, Mn, Zn and Cu are essential in trace quantities (Larcher, 1995). This section will be nearly exclusively devoted to nitrogen and phosphorus, as studies on nutrients in seagrasses are virtually confined to these elements. Levels of nitrogen and phosphorus in seagrass tissues have been most frequently reported for the leaves. Considerable variability exists in tissue levels. These vary between 0.5% and 5.5% of tissue dry weight for nitrogen and between 0.06% and 0.78% for phosphorus (Duarte, 1990, 1992). The frequency distributions of the nitrogen and phosphorus contents (Fig. 4.10) are positively skewed: the observed nutrient levels are concentrated in the lower end of the range, i.e. close to the minimum levels observed.

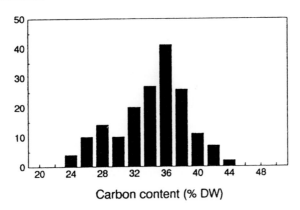

Carbon content (% DW)

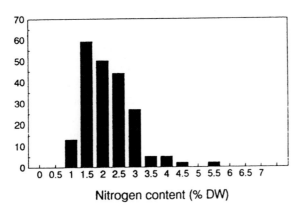

Nitrogen content (% DW)

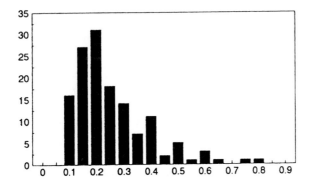

Phosphorus content (% DW)

Number of observations

Fig. 4.10. Frequency distribution of carbon, nitrogen, and phosphorus concentrations of seagrass leaves. (Duarte, 1990.)

Undoubtedly these lower limits are set by the fact that normal physiological cell functioning imposes a certain minimum content of nitrogen and phosphorus. The clustering of the majority of values in the low range of observed values may also suggest that nutrients are usually not accumulated at levels much above those required for normal cell functioning, but that there generally is a fairly close coupling between internal nutrient pools and leaf growth. However, the observed range in measured nitrogen and phosphorus content of seagrass leaves is considerably wider than that of carbon. Much of the latter element in seagrass leaves is part of cell wall structures, a rather invariable tissue component. The median carbon, nitrogen and phosphorus contents in seagrass leaves (33.5%, 1.9% and 0.24% of tissue dry weight, respectively; Duarte, 1992) translate into a median atomic C:N:P ratio of 435:20:1. In nitrogen limited situations, however, the C:N ratio may rise from 20 to above 40, whereas when phosphorus is severely limiting, C:P ratios may go up to 1000 or even higher.

The accepted atomic C:N:P ratio of phytoplankton cells in an unlimited nutrient environment is 106:16:1 (known as the Redfield ratio; Redfield *et al.*, 1963). Thus it can be seen that for biomass production on a carbon unit weight basis, seagrasses require approximately 4 times less nitrogen and phosphorus than phytoplankton cells. Moreover, the relative growth rate, i.e. the increase in tissue weight per unit of tissue weight per day, is typically much lower in seagrasses than in phytoplankton (Duarte, 1995). It thus follows that a hypothetical weight unit of seagrass requires approximately 50 times less nitrogen and 100 times less phosphorus for daily growth than a hypothetical weight unit of phytoplankton (Duarte, 1995). A similar calculation exercise shows that the nitrogen and phosphorus requirements for growth of macroalgae are about 8- and 1.5-fold higher, respectively, than those of seagrasses. These calculations are obviously only approximative and neglect differences between species. Nonetheless, they clearly suggest that seagrasses have an advantage for growth in nutrient-poor environments compared with other primary producers, because of their relatively low nutrient requirements.

4.4.2 *The nutrient environment of seagrasses*

Seagrasses use ammonium, nitrate and phosphate as ambient sources of nitrogen and phosphorus. These compounds are found in the water column and the sediment porewater. Nutrient levels in the water column of seagrass meadows are typically low, and particularly in tropical regions can be so low that accurate determinations are difficult. A compilation of data on the nutrient environment of seagrass meadows world-wide shows that the average water column ammonium concentration in seagrass beds is 3.1 µM, the nitrate concentration 2.7 µM, and the average phosphate concentration only 0.35 µM (Hemminga, 1998). Porewater nutrients are often present in much higher oncentrations. Ammonium in the porewater of seagrass sediments reaches a global average value of 86 µM, whereas phosphate on average is present at a concentration of 12 µM. Only the nitrate levels in the sediment are generally low, yielding an average value of 3.4 µM (Hemminga, 1998). This relatively low value can be ascribed to the usually anoxic character of the sediment, which restricts oxidation of ammonium formed during mineralization of organic matter.

Porewater nutrient levels show a large variability between sites, the particular concentrations at any site being the net result of regeneration and removal processes. Mineralization of organic matter is the major process responsible for the supply of inorganic nitrogen and phosphorus to the porewater. Fluxes of detritus to the pool of sediment organic matter and rates of microbial mineralization processes show pronounced local and regional variation, and, hence, explain part of the above variability in nutrient levels. Major removal processes are diffusion to the overlying water column, uptake by seagrasses and other organisms, and precipitation and adsorption. Adsorption of ammonium and phosphate to sediment particles occurs in all seagrass beds, but the importance of this process is strongly dependent on sediment characteristics. Hence, differences in sediment characteristics are another source of variability for porewater nutrient levels in seagrass meadows. This is well illustrated by the specific properties of carbonate sediments.

Carbonate-rich sediments are very often encountered in tropical and subtropical coastal waters. These sediments originate from the erosion of coral reefs and the fragmentation and accumulation of skeletal elements of molluscs, foraminifera, calcified algae, echinoderms, etc. Carbonate sediments contrast with siliceous sediments in having a high capacity for phosphate uptake (Kitano *et al.*, 1978). The mechanisms of uptake include adsorption of phosphate onto the sediment carbonate particles, and the precipitation of phosphate minerals such as apatite on their surface (Gaudette & Lyons, 1980). As a consequence the concentration of porewater phosphate in carbonate sediments is often very low. Fourqurean *et al.* (1992), for instance, measured levels of less than 0.5 µM in porewater of *Thalassia testudinum* meadows in Florida Bay. For comparison, ammonium levels were orders of magnitude higher, being around 100 μM. The resulting N:P ratio (ca. 200) is extremely high compared with that of the local seagrasses (ca. 25) which are expected to be the main source of organic matter available for decomposition in the sediment, and strongly suggests that much of the phosphate that is released during mineralization is bound to the carbonate matrix of the sediment, whereas this process is relatively less important for ammonium.

The dissolved phosphate levels in carbonate sediments are not always as low as in the above given example. The adsorption capacity of carbonates is related to the reactive surface area, which, in its turn, depends on the grain size of the sediment. This is demonstrated by an experiment of Erftemeyer & Middelburg (1993), who incubated a phosphate solution with carbonate sediment of an Indonesian *Thalassia testudinum* meadow. After three hours of incubation with sediment grains smaller than 0.075 mm, the initial phosphate concentration of 72 μM had dropped to 30 μM, whereas it had only decreased to 60 μM when grain size was between 0.6 and 1 mm. Grain-size dependent uptake of phosphate thus can explain the often higher phosphate porewater concentrations of coarse-grained sediments.

Apart from their potentially high phosphate uptake capacity, it can be mentioned here as an aside that carbonate sediments have another pecularity relevant in the context of nutrient limitation: these sediments bind iron as well. Iron is an essential micronutrient required for chlorophyll synthesis and a component of cytochromes. Tropical carbonate sediments often support seagrass vegetation that is deficient in iron. Adding iron to these sediments stimulated growth and increased the leaf chlorophyll concentration (Duarte *et al.*, 1995), providing the only evidence on the occurrence of nutrient limitation in natural seagrass vegetation other than that caused by a lack of available nitrogen and phosphorus.

4.4.3 Nutrient uptake

Laboratory studies have demonstrated that both leaves and the root/rhizome complex possess the capacity for nutrient uptake. Leaves readily take up ammonium, nitrate and phosphate when bathed in incubation media containing these compounds, whereas the root/rhizome complex has been shown to take up ammonium and phosphate. This double capacity for nutrient uptake is a general feature, and occurs in temperate, subtropical and tropical species alike. Uptake typically shows saturation kinetics, with the rate of uptake initially increasing in direct proportion to the external concentration, but levelling off and becoming saturated as ambient concentrations increase. On the basis of this hyperbolic pattern, the well-known Michaelis–Menten equation of enzyme kinetics is applied to describe the uptake rate in relation to the external concentration (Thursby & Harlin, 1982; Stapel *et al.*, 1996, Pedersen *et al.*, 1997; Terrados & Williams, 1997; Lee & Dunton, 1999; Fig. 4.11).

Seagrass leaves can be easily categorized into age, facilitating the assessment of uptake characteristics of individual age classes. Incubation experiments with radioactive and stable isotopes have demonstrated that the fully grown, mature leaves capture the lion's share of the inorganic nutrients. Mature leaves require few additional nutrients for growth, however, and therefore they mainly serve as conduits for nutrient supply to the young, growing leaves (Brix & Lyngby, 1985; Pedersen & Borum, 1992; Pedersen *et al.*, 1997; Stapel *et al.*, 1997). Detailed studies on the exact mechanism of nutrient uptake in seagrasses are lacking, but, as in other plants, transport across the cell membrane will be an active process, requiring the involvement of cellular metabolism for the provision of chemical energy.

The uptake capacities of seagrasses allow these plants to capture nutrients both from the water column and from the sediment. Although the often higher nutrient concentrations in the porewater would seem to make it obvious that the root system plays the dominant role in nutrient uptake, this is certainly not exclusively the case. In fact, the full range between the extremes of nutrient uptake dominated either by roots or by leaves can probably be found. Examples of the latter situation can be encountered in *Phyllospadix torreyi* and *Amphibolis antarctica*. These species can grow on rocky substrates, with no or little sediment around the roots. Laboratory studies show that the nitrogen uptake rates of the roots (on a unit dry weight basis) are only a fraction of leaf uptake rates in these plants (Pedersen *et al.*, 1997; Terrados & Williams, 1997). Apparently, the roots primarily serve to anchor them to the rocks.

Another example of nutrient uptake dominated by the leaves can be found in annual populations of *Zostera marina*. These plants develop from seeds in early spring, and may have a relatively high leaf to root biomass ratio. Repeated incubations of plants from a population in the south-west Netherlands during the growing season in locally collected seawater and sediment porewater showed that leaves dominated nitrogen uptake (mainly by absorption of ammonium, the most abundant

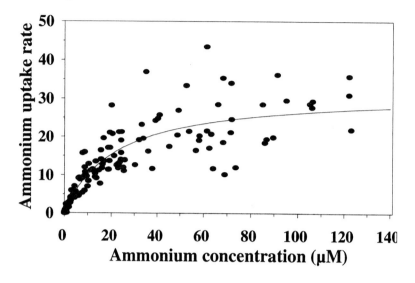

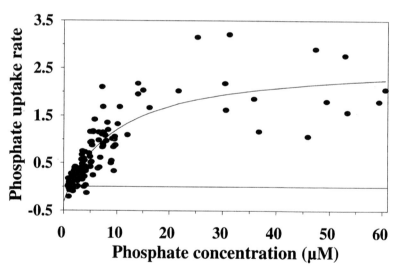

Fig. 4.11. Ammonium and phosphate uptake rates (μmol g^{-1} DW h^{-1}) by leaves of *Thalassia hemprichii* from Barang Lompo in the Spermonde Archipelago (Indonesia) as a function of ammonium and phosphate concentration in the incubation medium. The data were fitted to the Michaelis–Menten model to obtain the uptake curves shown. (Redrawn after Stapel *et al.*, 1996.)

inorganic nitrogen compound). Throughout the growing season, the leaves accounted for more than 70% of plant uptake, with a peak value of 92% in early September (Hemminga *et al.*, 1994). In perennial *Z. marina* the roots play a larger role in nitrogen uptake: determination of the annual dynamics of nitrogen in a Danish eelgrass population indicated that 49% of the nitrogen from the ambient media was captured by the leaves, and 51% by the roots (Pedersen & Borum, 1993).

Calculations on the relative importance of the leaves versus the roots for nutrient uptake based on uptake kinetics determined in laboratory incubations have their limitations. They give a short-term assessment of plant uptake characteristics, but even over the time scale of the experiment these may change. Transient enhanced uptake after exposure to a nutrient pulse, followed by decreasing uptake rates, have been documented (Pérez-Lloréns & Niell, 1995; Pedersen *et al.*, 1997). Uptake in the field, furthermore, is not only dependent on the ambient nutrient concentration, but also on flow rates, affecting the thickness of the diffusion boundary layer, and the nutrient supply rate. This may affect uptake by both the leaves and the roots. Stapel *et al.* (1996) calculated that if no mass flow of porewater occurred in the sediment of a *Thalassia hemprichii* vegetation, and the supply of nutrients thus would be dependent on the diffusion of nutrients from the porewater to the root surface, uptake would primarily be determined by diffusion limitation, and not by the uptake capacity of the roots. Other approaches (which will have their own limitations!) thus are very useful to supplement the data on root versus leaf uptake derived from *in vitro* studies of nutrient uptake. Erftemeyer & Middelburg (1995) applied a mass balance approach, using an extended data base on characteristics of reef-associated and coastal *Thalassia hemprichii* and *Enhalus acoroides* meadows, to define the contributions of leaves and roots in plant nutrient uptake. According to their model, root uptake could potentially account for 66–98% of annual nitrogen and phosphorus uptake in the absence of internal nutrient resorption. If the requirements for nitrogen and phosphorus uptake from external sources were partially met by nutrient resorption, the importance of leaf uptake increased, contributing approximately half of total plant uptake at levels of 50% of internal nitrogen and phosphorus resorption. These results are consistent with the picture emerging from laboratory studies to the extent that they show that nutrient uptake in seagrasses occurs over the entire plant surface, the proportional contributions of aboveground and belowground organs being dependent on local environment conditions and plant properties.

4.4.4 Nutrient limitation

Studies of nutrient limitation in seagrasses are routinely carried out by enrichment of the sediment with inorganic nitrogen- and phosphorus-containing compounds (e.g. by introduction of commercially available slow-release nutrient pellets), followed by growth analyses. The reasons why virtually all studies of natural systems apply sediment fertilization are threefold: historically, the roots have been considered as the main organ for nutrient uptake; furthermore, sediment fertilization is more easily accomplished than nutrient enrichment of the water column in natural systems; and, finally, enrichment of the water column may have simultaneous effects on other primary producers such as macroalgae and epiphytic leaf algae (Harlin & Thorne-Miller, 1981; Wear *et al.*, 1999), which may complicate the interpretation of the results.

Increased growth of seagrasses after sediment fertilization has been observed many times, suggesting that biomass formation is often limited by nutrient availability. Nutrient limitation involves only one nutrient at a time. Addition of this nutrient can be followed by growth limitation by the next nutrient that is in short supply. This explains observations of elevated growth in treatments with combined nutrient additions (N + P) relative to growth stimulation brought about by supply of the primary limiting nutrient only. In exceptional cases, the deficiency of two nutrients can be so closely balanced that only simultaneous addition results in stimulated growth (Udy & Dennison, 1997). Seagrasses growing on carbonate sediments have repeatedly been demonstrated to be phosphorus limited (Short *et al.*, 1990; Pérez *et al.*, 1991; Fourqurean *et al.*, 1992), a phenomenon that can be related to the phosphate-binding characteristics of carbonate sediments, discussed above. Recently, evidence has been found that in the seagrass rooting zone the opposite process may also occur: the release of phosphorus from the carbonate matrix (Jensen *et al.*, 1998). Such a mobilization of phosphorus probably results from dissolution of carbonate in the rhizosphere due to a lowered pH or to the direct action of organic acids released by the roots. The excretion of acidifying and chelating compounds that facilitate the acquisition of soil phosphate is well documented for roots of a number of terrestrial plants (Lambers *et al.*, 1998), and thus is perhaps also an option for seagrasses to alleviate phosphorus limitation in carbonate soils.

Although phosphorus limitation is typically observed in meadows growing on carbonate sediments, it may also occur in vegetation on siliceous sediments. The available data indicate that if nutrient limitation occurs in the latter type of sediments, it concerns either a combination of nitrogen and phosphorus limitation, or nitrogen limitation exclusively (Bulthuis *et al.*, 1992; Murray *et al.*, 1992; Short *et al.*, 1993; Udy & Dennison, 1997).

Nutrient-limited growth may be a seasonally transient phenomenon, or have a more permanent character. Seasonal nutrient limitation is more likely to occur in temperate than in tropical meadows, where permanent nutrient limitation is more probable. In contrast to the rather stable light and temperature conditions of the tropics, the pronounced seasonality in temperate regions coincides with restricted periods of fast growth. In these regions, nutrient limitation is particularly likely to occur in late spring and summer, when water temperatures have been rising after the winter minimum, and irradiance levels are high. Hence, a shift from growth control by temperature and light during winter to nutrient-limited growth in summer may occur (Alcoverro *et al.*, 1997). Nutrient limitation also is not necessarily identical for all species growing in the same area. Udy & Dennison (1997) showed that in an area where *Halodule uninervis*, *Zostera capricorni* and *Cymodocea serrulata* occurred in adjacent monospecific beds, growth of the first species was nitrogen limited, that of the second was limited by a balanced deficiency of both nitrogen and phosphorus, whereas growth of *C. serrulata* did not appear to be nutrient limited at all (Fig. 4.12). Such differences may arise from species-specific nutrient requirements or physiological capacities for nutrient uptake, but it is also possible that different rooting depths, leading to the exploration of nutrient pools in different sediment layers, play a role (Agawin *et al.*, 1996).

The interplay of many abiotic and plant-related factors, which ultimately determine whether seagrass growth is nutrient limited or not, render statements about the occurrence of nutrient limitation at a given site uncertain in the absence of laborious and time-consuming fertilization experiments. Tissue nutrient contents are generally considered indicative for nutrient availability in seagrasses and other aquatic angiosperms. For seagrasses, the median nitrogen and phosphorus concentrations established in seagrass leaves (1.8% and 0.2%, respectively; Duarte, 1990), can be taken as reference points for assessing the likelihood of nutrient limitation. Data so far indicate that there is a fair chance that tissue contents considerably below these values reflect nutrient limitation. It should be borne in mind, however, that uniform critical tissue nutrient levels generally applicable to seagrasses as a group

Halodule

Zostera

Cymodocea

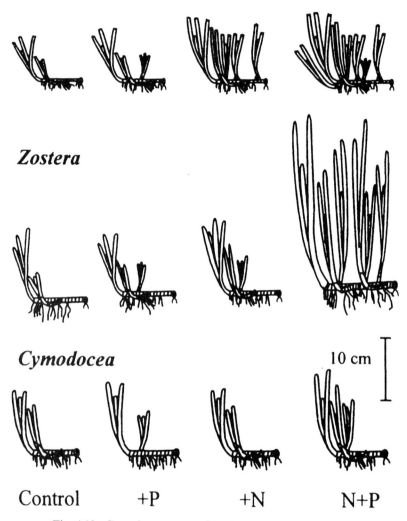

Control   +P   +N   N+P

Fig. 4.12. Growth response of *Halodule uninervis, Zostera capricorni* and *Cymodocea serrulata* to fertilization with phosphorus (+P), nitrogen (+N) and nitrogen plus phosphorus (N+P) Height of the leaves and number of shoots are indicative of the growth response. (Udy & Dennison, 1997.)

probably do not exist. Research carried out in more recent years has shown that growth limitation may occur simultaneously with nutrient levels higher than median, whereas in other situations nutrient limitation could not be demonstrated despite low tissue nutrient levels (Erftemeyer *et al.*, 1994; Pedersen, 1995; Agawin *et al.*, 1996; Udy & Dennison, 1997).

4.4.5 *Effects of nutrient limitation*

Nutrient limitation affects the functioning of individual shoots and clones, and generally results in reduced leaf productivity and areal biomass. It often also coincides with a relatively low shoot density. Enhanced photosynthetic capacity (on basis of leaf dry weight units), following sediment nutrient additions, has been observed in several Philippine seagrasses (Agawin *et al.*, 1996), suggesting that nutrient limitation may lead to a decreased photosynthetic performance. There is also some evidence that low nutrient availability leads to a relatively larger allocation of biomass to the roots than to the shoots (Short *et al.*, 1985; Pérez *et al.*, 1994), as in terrestrial plants. It is likely that the plants thereby increase their capacity to acquire nutrients from the sediment pool. This response is not self-evident in seagrasses considering their leaf nutrient uptake capacity. It thus appears that the growth strategies of seagrasses facing either light or nutrient limitation resemble those of terrestrial plants, with light limitation shifting biomass allocation to the leaf compartment (see Fig. 4.7), and nutrient limitation leading to an emphasis on root growth.

In multi-species systems, the effects of nutrient availability on growth of individual species can be expected to have an impact on plant community composition, including features such as the dominance or exclusion of species. This can be derived from observations on *Thalassia testudinum* and *Halodule wrightii* dominated seagrass meadows in Florida Bay. These meadows occur on carbonate sediments, and their growth is phosphorus limited as is demonstrated by the fact that seagrass standing crop is positively correlated with the concentration of dissolved inorganic phosphorus in the sediment porewater. The porewater phosphate concentration in the areas supporting *H. wrightii* was higher than in areas with *T. testudinum* (Fourqurean *et al.*, 1992). In the *T. testudinum* meadows, an elegant long-term fertilization experiment was carried out with birds employed as a kind of cheap labour force. By placing poles in the sediment of the meadows, continuous fertilization of the sites by birds perched on the poles could be brought about. The poles were

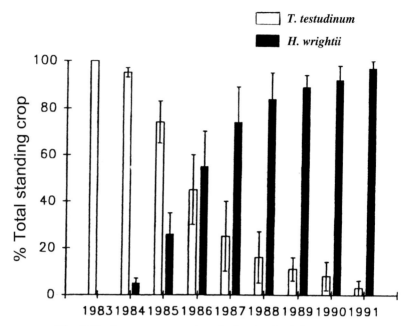

Fig. 4.13. Average proportion (± 1 SE) of *Thalassia testudinum* and *Halodule wrightii* at long-term fertilized sites. (Fourqurean *et al.*, 1995.)

heavily used as roosting sites by terns and cormorants, and this coincided with continuous deposition of nutrient-rich faeces in the local beds. The standing crop of *T. testudinum* doubled in the two years after the start of fertilization. During the first years of the observations, however, *H. wrightii* also entered the fertilized patches, one patch after the other. After *H. wrightii* became established, the standing crop of *T. testudinum* decreased, and after 8 years the fertilized sites were completely dominated by *H. wrightii* (Fourqurean *et al.*, 1995; Fig. 4.13). The dominance of *T. testudinum* in the unfertilized situation is probably related to the approximately four times lower phosphorus demand of this species compared with *H. wrightii*, which allows it to outcompete *H. wrightii* in an environment where phosphate is in short supply. Increased nutrient levels following fertilization allowed *H. wrightii* to expand its biomass and, ultimately, to completely displace *T. testudinum*.

4.4.6 Nutrient conservation in seagrasses

Although we do not know how frequently nutrient limitation of seagrass growth occurs, the abundant literature on this topic suggests that it certainly is not a rare phenomenon. Seagrasses, in general, thus would profit from mechanisms that would diminish their dependence on uptake of nutrients from the environment. The relatively low tissue nutrient levels of seagrasses help to restrict the demands for external nutrients. Two other plant characteristics are potentially very important in the reduction of the amount of nutrients that must be captured from the environment: nutrient reuse and leaf longevity. Unlike carbon, an element that for a large part is fixed in the lignine–cellulose complex of cell walls, nitrogen and phosphorus are particularly found as constituents of compounds that actively participate in cell metabolism. The reuse of nitrogen and phosphorus within the plant structure, by resorption of these nutrients from senescing tissue and reallocation to young growing tissues, is a common phenomenon in vascular plants. This process is presumed to be particularly important for plants growing in nutrient-poor environments, and has been studied in the leaves of numerous (mainly terrestrial) plants.

The nutrient resorption efficiency of plant leaves is defined as the amount of nutrients resorbed during senescence and is expressed as the percentage reduction of nutrients between mature green and senescent leaves (Aerts, 1996). A compilation of data published in seagrass studies shows that seagrasses, on average, resorb 20.4% of the leaf nitrogen, and 21.9% of the leaf phosphorus (Hemminga *et al.*, 1999). These average figures concern resorption from undamaged leaves. The hydrodynamic stresses imposed upon the leaves, however, which may lead to premature losses of leaf fragments, and the grazing activity of herbivorous animals, may reduce the amount that is actually resorbed to a lower value than is physiologically possible in undamaged leaves (Stapel & Hemminga, 1997). This aspect has not been included in the average figures on nutrient resorption cited above, but even so the resorption efficiencies of seagrasses are very low compared with those of perennial terrestrial plants (Table 4.1), where nitrogen and phosphorus resorption varies between 41% and 58% (nitrogen) and between 42% and 71% (phosphorus). Nutrient resorption from senescing leaves apparently is not a pronounced conservation strategy in seagrasses.

An extended leaf longevity would be an alternative way of reducing nutrient demands, since this lowers the frequency of leaf formation and extends the time available to reclaim nutrients from mature leaves. However, again seagrasses compare unfavourably with terrestrial plants. Seagrass leaf longevity is highly variable, the values ranging between 345 days in *Posidonia oceanica* to 4 days in *Halophila ovalis*, with a mean of

Table 4.1. *Nitrogen and phosphorus resorption efficiency (%R_N and %R_P, respectively) in terrestrial perennial plants and in seagrasses*

Means ± SD. The number of observations are shown in parentheses

Plant group	$\%R_N$	$\%R_P$
Evergreen shrubs and trees	46.7 ± 16.4 (108)	51.4 ± 21.7 (88)
Deciduous shrubs and trees	54.0 ± 15.9 (115)	50.4 ± 19.7 (98)
Forbs	41.4 ± 21.4 (33)	42.4 ± 30.3 (18)
Graminoids	58.5 ± 14.2 (31)	71.5 ± 16.0 (22)
Seagrasses	**20.4 ± 12.2 (19)**	**21.9 ± 17.0 (12)**

Source: Data on terrestrial plants are derived from Aerts (1996). The figures on seagrasses are derived from Hemminga *et al.* (1999)

88 days; as a comparison, the average leaf life of terrestrial herbs is just over 170 days (Hemminga *et al.*, 1999). It is plausible that the relatively short leaf life of seagrasses is related to the gradual coverage of the leaf surfaces with epiphytes and silt that normally accompanies ageing. This will impede photosynthesis and perhaps nutrient uptake as well, without affecting the maintenance costs of the leaves. From that perspective, the possession of long-lived leaves thus is not advantageous. The limited importance of leaf resorption, and the fact that leaf longevity is relatively short, implies that seagrass plants lose nutrients at high rates, and, thus, have to rely more strongly on uptake of nutrients from the environment than many terrestrial angiosperms do (Hemminga *et al.*, 1999). The 'double' uptake potential of seagrasses, via both roots and leaves, may be an advantage in that respect.

It should be added that the clonal architecture of seagrasses provides them with the means of sharing resources between ramets, a feature that may alleviate the nutrient stress of individual ramets. As the ramets are spatially separated, they may experience different conditions of nutrient availability. As long as connections are intact, the seagrass clone may function as a physiologically integrated system in which nutrients are transported from one point to another. The existence and functional importance of source–sink relations between interconnected shoots is a largely unexplored area in the field of seagrass research. Transport of carbon between shoots does occur, as experiments with ^{14}C indicate (Libes & Boudouresque, 1987), and undoubtedly explains the unhampered functioning of seagrass shoots that are kept in the shade but which are physically connected to neighbouring non-shaded shoots (Tomasko & Dawes, 1989). Likewise, nutrient fluxes between shoots exist, as is demonstrated by ^{15}N-tracer experiments. In these experiments individual shoots of a number of different seagrass species were enclosed within transparent bags containing ^{15}N-labelled ammonium in seawater. These incubations were carried out in natural meadows. Four days after this initial incubation, all shoots along the rhizome were harvested. Isotope analysis showed that transport of nitrogen to neighbouring shoots occurs in all species, even to shoots that are separated from the incubated shoots by more than 10 other shoots (N. Marbà, M. Hemminga & M. Mateo, unpublished results). These findings are consistent with observations of Terrados *et al.* (1997), who reported that severing the horizontal rhizome of *Cymodocea nodosa* reduced biomass production by the apical meristem and leaf growth of the shoots on the isolated rhizome fragment, even in rhizome pieces with 11 shoots. Available resources as nutrients and probably carbon compounds thus can be transported a considerable distance away from the uptake or production site, to support growth of distant ramets. An example of the advantages clonal integration produces in nutrient-limited situations has also been described in *C. nodosa*, where resource sharing supports growth of young shoots at colonizing rhizome apices in phosphorus-deficient sediments, whereas seedlings fail to grow (Duarte & Sand-Jensen, 1996).

4.5 Concluding remarks

Seagrasses have the same basic requirements for growth as terrestrial angiosperms: light, inorganic carbon and nutrients. The availability of these resources, however, is frequently restricted, a fact that may be exacerbated by ecophysiological and structural characteristics of the plants. Light penetrates with difficulty in water, and the presence of suspended particles and of various kinds of dissolved substances aggravates this feature. At already shallow depths, therefore, only a fraction of the surface irradiance is left. Inorganic carbon is present in considerable amounts in seawater, but evidence of efficient direct uptake of bicarbonate ions, as found in algae, is still scanty. Finally, the nutrient uptake capacity of both leaves and roots is advantageous for the acquisition of nutrients, but the physiological efficiency of nutrient use by seagrasses compares unfavourably with that of many terrestrial angiosperms.

It is thus not surprising that growth of seagrasses is often suboptimal as a consequence of resource limitation. Such limitations may vary in time and place, and may shift from one resource to another. To some extent these limitations are predictable. With increasing water depth, light limitation becomes inevitable. Light limitation at very shallow depths is expected in and near estuaries and river plumes that are loaded with terrestrial particles, and in nutrient-rich waters that allow extensive phytoplankton productivity. In areas with limited land run-off, transparency of the water generally will be high, but such areas also lack the supply of nutrients coinciding with land run-off. Shallow meadows growing there are thus more likely to experience nutrient limitation than light limitation. Finally, resource limitation generally will have a more pronounced seasonal character in temperate regions than in tropical regions, because the low winter temperatures in temperate regions result in temporarily diminished resource requirements for growth. Seasonally changing environments also often imply a succession of resources that become limiting one after the other, related to the different cycles in their availability.

4.6 References

Abel, K. (1984). Inorganic carbon source for photosynthesis in the seagrass *Thalassia hemprichii* (Ehrenb.) Aschers. *Plant Physiology*, **76**, 776–81.

Aerts, R. (1996). Nutrient resorption from senescing leaves of perennials: are there general patterns? *Journal of Ecology*, **84**, 597–608.

Agawin, N.S.R., Duarte, C.M. & Fortes, M.D. (1996). Nutrient limitation of Philippine seagrasses (Cape Bolinao, NW Philippines): in situ experimental evidence. *Marine Ecology Progress Series*, **138**, 233–43.

Agustí, S., Enríquez, S., Frost-Christensen, H., Sand-Jensen, K. & Duarte, C.M. (1994). Light harvesting among photosynthetic organisms. *Functional Ecology*, **8**, 273–79.

Alcoverro, T., Romero, J., Duarte, C.M. & López, N.I. (1997). Spatial and temporal variations in nutrient limitation of seagrass *Posidonia oceanica* growth in the NW Mediterranean. *Marine Ecology Progress Series*, **146**, 155–61.

Alcoverro, T., Manzanera, M. & Romero, J. (1998). Seasonal and age-dependent variability of *Posidonia oceanica* (L.) Delile photosynthetic parameters. *Journal of Experimental Marine Biology and Ecology*, **230**, 1–13.

Beer, S. (1989). Photosynthesis and photorespiration of marine angiosperms. *Aquatic Botany*, **43**, 153–66.

Beer, S. & Koch, E. (1996). Photosynthesis of seagrasses vs. marine macroalgae in globally changing CO_2 environments. *Marine Ecology Progress Series*, **141**, 199–204.

Beer, S. & Rehnberg, J. (1997). The acquisition of inorganic carbon by the seagrass *Zostera marina*. *Aquatic Botany*, **56**, 277–83.

Beer, S. & Waisel, Y. (1979). Some photosynthetic carbon fixation properties of seagrasses. *Aquatic Botany*, **7**, 129–38.

Beer, S. & Waisel, Y. (1982). Effects of light and pressure on photosynthesis in two seagrasses. *Aquatic Botany*, **13**, 331–7.

Beer, S., Vilenkin, B., Weil, A., Veste, M., Susel, L. & Eshel, A. (1998).

Measuring photosynthetic rates in seagrasses by pulse amplitude modulated (PAM) fluorometry. *Marine Ecology Progress Series*, **174**, 293–300.

Björk, M., Weil, A., Semesi, S. & Beer, S. (1997). Photosynthetic utilisation of inorganic carbon by seagrasses from Zanzibar, East Africa. *Marine Biology*, **129**, 363–6.

Brix, H. & Lyngby, J.E. (1985). Uptake and translocation of phosphorus in eelgrass (*Zostera marina*). *Marine Biology*, **90**, 111–16.

Brouwer, R. (1962). Nutritive influences on distribution of dry matter in the plant. *Netherlands Journal of Agricultural Science*, **10**, 399–408.

Brouwer, R. (1963). Some aspects of the equilibrium between overground and underground plant parts. *Mededelingen van het Instituut voor Biologisch en Scheikundig Onderzoek van Landbouw Gewassen*, 213, **31–9**.

Bulthuis, D.A., Axelrad, D.M. & Mickelson, M.J. (1992). Growth of the seagrass *Heterozostera tasmanica* limited by nitrogen in Port Phillip Bay, Australia. *Marine Ecology Progress Series*, **89**, 269–75.

Burke, M.K., Dennison, W.C. & Moore, K.A. (1996). Non-structural carbohydrate reserves of eelgrass *Zostera marina*. *Marine Ecology Progress Series*, **137**, 195–201.

Cloern, J.E. (1996). Phytoplankton bloom dynamics in coastal ecosystems: a review with some general lessons from sustained investigation of San Francisco Bay, California. *Reviews of Geophysics*, **34**, 127–68.

Connell, E.L., Colmer, T.D. & Walker, D.I. (1999). Radial oxygen loss from intact roots of *Halophila ovalis* as a function of distance behind the root tip and shoot illumination. *Aquatic Botany*, **63**, 219–28.

Czerny, A.B. & Dunton, K.H. (1995). The effects of in situ light reduction on the growth of two subtropical seagrasses *Thalassia testudinum* and *Halodule wrightii*. *Estuaries*, **18**, 418–27.

Dawes, C.J. (1998). Biomass and photosynthetic responses to irradiance by a shallow and a deep water population of *Thalassia testudinum* on the west coast of Florida. *Bulletin of Marine Science*, **62**, 89–96.

Den Hartog, C. (1970). *The Seagrasses of the World*. Amsterdam: North Holland Publishing Company.

Dennison, W.C. & Alberte, R.S. (1982). Photosynthetic responses of *Zostera marina* L. (eelgrass) to in situ manipulations of light intensity. *Oecologia*, **55**, 137–44.

Dennison, W.C. & Alberte, R.S. (1985). Role of daily light period in the depth distribution of *Zostera marina* (eelgrass). *Marine Ecology Progress Series*, **25**, 51–61.

Dennison, W.C. & Alberte, R.S. (1986). Photoadaptation and growth of *Zostera marina* L. (eelgrass) transplants along a depth gradient. *Journal of Experimental Marine Biology and Ecology*, **98**, 265–82.

Drew, E.A. (1979). Physiological aspects of primary production in seagrasses. *Aquatic Botany*, **7**, 139–50.

Dring, M.J. (1982). *The Biology of Marine Plants*. Cambridge: Cambridge University Press.

Duarte, C.M. (1989). Temporal biomass variability and production/biomass relationships of seagrass communities. *Marine Ecology Progress Series*, **51**, 269–76.

Duarte, C.M. (1990). Seagrass nutrient content. *Marine Ecology Progress Series*, **67**, 201–7.

Duarte, C.M. (1991). Seagrass depth limits. *Aquatic Botany*, **40**, 363–77.

Duarte, C.M. (1992). Nutrient concentrations of aquatic plants: patterns across species. *Limnology and Oceanography*, **37**, 882–9.

Duarte, C.M. (1995). Submerged aquatic vegetation in relation to different nutrient regimes. *Ophelia*, **41**, 87–112.

Duarte, C.M. & Sand-Jensen, K. (1996). Nutrient constraints on establishment from seed and on vegetative expansion of the Mediterranean seagrass *Cymodocea nodosa*. *Aquatic Botany*, **54**, 279–86.

Duarte, C.M., Merino, M. & Gallegos, M. (1995). Evidence of iron deficiency in seagrasses growing above carbonate sediments. *Limnology and Oceanography*, **40**, 1153–8.

Dunton, K.H. (1994). Seasonal growth and biomass of the subtropical seagrass *Halodule wrightii* in relation to continuous measurements of underwater irradiance. *Marine Biology*, **120**, 479–89.

Dunton, K.H. & Tomasko, D.A. (1994). In situ photosynthesis in the seagrass *Halodule wrightii* in a hypersaline subtropical lagoon. *Marine Ecology Progress Series*, **107**, 281–93.

Durako, M.J. (1993). Photosynthetic utilization of $CO_2(aq)$ and HCO_3^- in *Thalassia testudinum* (Hydrocharitaceae). *Marine Biology*, **115**, 373–80.

Enríquez, S., Duarte, C.M. & Sand-Jensen, K. (1995). Patterns in the photosynthetic metabolism of Mediterranean macrophytes. *Marine Ecology Progress Series*, **119**, 243–52.

Erftemeijer, P.L.A. & Herman, P.M.J. (1994). Seasonal changes in environmental variables, biomass, production and nutrient contents in two contrasting tropical intertidal seagrass beds in South Sulawesi, Indonesia. *Oecologia*, **99**, 45–59.

Erftemeijer, P.L.A. & Middelburg, J.J. (1993). Sediment–nutrient interactions in tropical seagrass beds: a comparison between a terrigenous and a carbonate sedimentary environment in south Sulawesi (Indonesia). *Marine Ecology Progress Series*, **102**, 187–98.

Erftemeijer, P.L.A. & Middelburg, J.J. (1995). Mass balance constraints on nutrient cycling in tropical seagrass beds. *Aquatic Botany*, **50**, 21–36.

Erftemeijer, P.L.A., Stapel, J., Smekens, M.J.E. & Drossaert, W.M.E. (1994). The limited effect of in situ phosphorus and nitrogen additions to seagrass beds on carbonate and terrigenous sediments in South Sulawesi, Indonesia. *Journal of Experimental Marine Biology and Ecology*, **182**, 123–40.

Farquhar, G.D., Ehleringer, J.R., Hubick, K.T. (1989). Carbon isotope discrimination and photosynthesis. *Annual Review of Plant Physiology and Plant Molecular Biology*, **40**, 503–37.

Fitzpatrick, J. & Kirkman, H. (1995). Effects of prolonged shading stress on growth and survival of seagrass *Posidonia australis* in Jervis Bay, New South Wales, Australia. *Marine Ecology Progress Series*, **127**, 279–89.

Fourqurean, J.W. & Zieman, J.C. (1991). Photosynthesis, respiration and whole plant carbon budget of the seagrass *Thalassia testudinum*. *Marine Ecology Progress Series*, **69**, 161–70.

Fourqurean, J.W., Zieman, J.C. & Powell, G.V.N. (1992). Relationships between porewater nutrients and seagrasses in a subtropical carbonate environment. *Marine Biology*, **114**, 57–65.

Fourqurean, J.W., Powell, G.V.N., Kenworthy, W.J. & Zieman, J.C. (1995). The effects of long-term manipulation of nutrient supply on competition between the seagrasses *Thalassia testudinum* and *Halodule wrightii* in Florida Bay. *Oikos*, **72**, 349–58.

Gaudette, H.E. & Lyons, W.B. (1980). Phosphate geochemistry in nearshore carbonate sediments: a suggestion of apatite formation. *SEPM Special Publication*, 29, **215–25**.

Gordon, D.M., Grey, K.A. & Simpson, C.J. (1994). Changes to the structure and productivity of a *Posidonia sinuosa* meadow during and after imposed shading. *Aquatic Botany*, **47**, 265–75.

Grice, A.M., Loneragan, N.R. & Dennison, W.C. (1996). Light intensity and the interactions between physiology, morphology and stable isotope ratios in five species of seagrass. *Journal of Experimental Marine Biology and Ecology*, **195**, 91–110.

Harlin, M.M. & Thorne-Miller, B. (1981). Nutrient enrichment of seagrass beds in a Rhode Island coastal lagoon. *Marine Biology*, **65**, 221–9.

Hellblom, F. & Björk, M. (1999). Photosynthetic responses in *Zostera marina* to decreasing salinity, inorganic carbon content and osmolality. *Aquatic Botany*, **65**, 97–104.

Hemminga, M.A. (1998). The root/rhizome system of seagrasses: an asset and a burden. *Journal of Sea Research*, **39**, 183–96.

Hemminga, M.A. & Mateo, M.A. (1996). Stable carbon isotopes in seagrasses: variability in ratios and use in ecological studies. *Marine Ecology Progress Series*, **140**, 285–98.

Hemminga, M.A., Koutstaal, B.P., Van Soelen, J. & Merks, A.G.A. (1994). The nitrogen supply to intertidal eelgrass (*Zostera marina*). *Marine Biology*, **118**, 223–7.

Hemminga, M.A., Marbà, N. & Stapel, J. (1999). Leaf nutrient resorption, leaf lifespan and the retention of nutrients in seagrass systems. *Aquatic Botany*, **65**, 141–58.

Herzka, S.Z. & Dunton, K.H. (1997). Seasonal photosynthetic patterns of the seagrass *Thalassia testudinum* in the western Gulf of Mexico. *Marine Ecology Progress Series*, **152**, 103–17.

Hillman, K., Walker, D.I., Larkum, A.W.D. & McComb, A.J. (1989). Productivity and nutrient limitation. In *Biology of Seagrasses*, ed. A.W.D. Larkum, A.J. McComb & S.A. Shepherd, pp. 635–685. Amsterdam: Elsevier.

Invers, O., Romero, J. & Pérez, M. (1997). Effects of pH on seagrass photosynthesis: a laboratory and field assessment. *Aquatic Botany*, **59**, 185–94.

Invers, O., Pérez, M. & Romero, J. (1999). Bicarbonate utilization in seagrass photosynthesis: role of carbonic anhydrase in *Posidonia oceanica* (L.) Delile and *Cymodocea nodosa* (Ucria) Ascherson. *Journal of Experimental Marine Biology and Ecology*, **235**, 125–33.

James, P.L. & Larkum, A.W.D. (1996). Photosynthetic inorganic carbon acquisition of *Posidonia australis*. *Aquatic Botany*, **55**, 149–57.

Jensen, H.S., McGlathery, K.J., Marino, R. & Howarth, R.W. (1998). Forms and availability of sediment phosphorus in carbonate sand of Bermuda seagrass beds. *Limnology and Oceanography*, **43**, 799–810.

Kenworthy, W.J. & Fonseca, M.S. (1996). Light requirements of seagrasses *Halodule wrightii* and *Syringodium filiforme* derived from the relationship between diffuse light attenuation and maximum depth distribution. *Estuaries*, **19**, 740–50.

Kirk, J.T.O. (1983). *Light and Photosynthesis in Aquatic Ecosystems*. Cambridge: Cambridge University Press.

Kitano, Y., Okumura, M. & Idogak, M. (1978). Uptake of phosphate ions by calcium carbonate. *Geochemical Journal*, **12**, 29–37.

Koch, E.W. (1994). Hydrodynamics, diffusion-boundary layers and photosynthesis of the seagrasses *Thalassia testudinum* and *Cymodocea nodosa*. *Marine Biology*, **118**, 767–76.

Kraemer, G.P. & Alberte, R.S. (1993). Age-related patterns of metabolism and biomass in subterranean tissues of *Zostera marina* (eelgrass). *Marine Ecology Progress Series*, **95**, 193–203.

Krause, G.H. & Weis, E. (1991). Chlorophyll fluorescence and photosynthesis: the basics. *Annual Review of Plant Physiology and Plant Biology*, **42**, 313–49.

Lambers, H., Chapin III, F.S. & Pons, T.L. (1998). *Plant Physiological Ecology*. Berlin: Springer.

Lanyon, J.M. & Marsh, H. (1995). Temporal changes in the abundance of some tropical intertidal seagrasses in North Queensland. *Aquatic Botany*, **49**, 217–37.

Larcher, W. (1995). *Physiological Plant Ecology*. Berlin: Springer.

Larkum, A.W.D., Roberts, G., Kuo, J. & Strother, S. (1989). Gaseous movement in seagrasses. In *Biology of Seagrasses*, ed. A.W.D. Larkum, A.J. McComb & S.A. Shepherd, pp. 686–722. Amsterdam: Elsevier.

Laugier, T., Rigollet, V. & De Casabianca, M.-L. (1999). Seasonal dynamics in mixed eelgrass beds, *Zostera marina* L. and *Z. noltii* Hornem., in a Mediterranean coastal lagoon (Thau lagoon, France). *Aquatic Botany*, **63**, 51–69.

Lee, K.-S.& Dunton, K.H. (1997). Effects of in situ light reduction on the maintenance, growth and partitioning of carbon resources in *Thalassia testudinum* Banks ex König. *Journal of Experimental Marine Biology and Ecology*, **210**, 53–73.

Lee, K.-S.& Dunton, K.H. (1999). Inorganic nitrogen acquisition in the seagrass *Thalassia testudinum*: development of a whole-plant nitrogen budget. *Limnology and Oceanography*, **44**, 1204–15.

Leuschner, C. & Rees, U. (1993). CO_2 gas exchange of two intertidal seagrass species, *Zostera marina* L. and *Zostera noltii* Hornem., during emersion. *Aquatic Botany*, **45**, 53–62.

Libes, M. & Boudouresque, C.F. (1987). Uptake and long-distance transport of carbon in the marine phanerogam *Posidonia oceanica*. *Marine Ecology Progress Series*, **38**, 177–86.

Longstaff, B.J., Loneragan, N.R., O'Donohue, M.J. & Dennison, W.C. (1999). Effects of light deprivation on the survival and recovery of the seagrass *Halophila ovalis* (R.Br.) Hook. *Journal of Experimental Marine Biology and Ecology*, **234**, 1–27.

Maberly, S.C., Raven, J.A., Johnston, A.M. (1992). Discrimination between ^{12}C and ^{13}C by marine plants. *Oecologia*, **91**, 481–92.

Marbà, N., Cebrián, J., Enríquez, S. & Duarte, C.M. (1996). Growth patterns of Western Mediterranean seagrasses: species-specific responses to seasonal forcing. *Marine Ecology Progress Series*, **133**, 203–15.

Masini, R.J., Manning, C.R. (1997). The photosynthetic responses to irradiance and temperature of four meadow-forming seagrasses. *Aquatic Botany* **58**, 21–36.

Mellors, J.E., Marsh, H. & Coles, R.G. (1993). Intra-annual changes in seagrass standing crop, Green Island, Northern Queensland. *Australian Journal of Marine and Freshwater Research*, **44**, 33–41.

Murray, L., Dennison, W.C. & Kemp, W.M. (1992). Nitrogen versus phosphorus limitation for growth of an estuarine population of eelgrass (*Zostera marina* L.). *Aquatic Botany*, **44**, 83–100.

Nielsen, S.L. & Sand-Jensen, K. (1989). Regulation of photosynthetic rates of submerged rooted macrophytes. *Oecologia*, **81**, 364–8.

Olesen, B. & Sand-Jensen, K. (1993). Seasonal acclimatization of eelgrass *Zostera marina* growth to light. *Marine Ecology Progress Series*, **94**, 91–9.

Orth, R.J. & Moore, K.A. (1986). Seasonal and year-to-year variations in the growth of *Zostera marina* L. (eelgrass) in the lower Chesapeake Bay. *Aquatic Botany*, **24**, 335–41.

Paneth, P. & O'Leary, M.H. (1985). Carbon isotope effect on dehydration of bicarbonate ion catalyzed by carbonic anhydrase. *Biochemistry*, **24**, 5143–7.

Pedersen, M.F. (1995). Nitrogen limitation of photosynthesis and growth: comparison across aquatic plant communities in a Danish estuary (Roskilde Fjord). *Ophelia*, **41**, 261–72.

Pedersen, M.F. & Borum, J. (1992). Nitrogen dynamics of eelgrass *Zostera marina* during a late summer period of high growth and low nutrient availability. *Marine Ecology Progress Series*, **80**, 65–73.

Pedersen, M.F. & Borum, J. (1993). An annual nitrogen budget for a seagrass *Zostera marina* population. *Marine Ecology Progress Series*, **101**, 169–77.

Pedersen, M.F., Paling, E.I. & Walker, D.I. (1997). Nitrogen uptake and allocation in the seagrass *Amphibolis antarctica*. *Aquatic Botany*, **56**, 105–17.

Pedersen, O., Borum, J., Duarte, C.M. & Fortes, M.D. (1998). Oxygen dynamics in the rhizosphere of *Cymodocea rotundata*. *Marine Ecology Progress Series*, **169**, 283–8.

Penhale, P.A. & Wetzel, R.G. (1983). Structural and functional adaptations of eelgrass (*Zostera marina*) to the anaerobic sediment environment. *Canadian Journal of Botany*, **61**, 1421–8.

Pérez, M. & Romero, J. (1992). Photosynthetic response to light and temperature of the seagrass *Cymodocea nodosa* and the prediction of its seasonality. *Aquatic Botany*, **43**, 51–62.

Pérez, M., Romero, J., Duarte, C.M. & Sand-Jensen, K. (1991). Phosphorus limitation of *Cymodocea nodosa* growth. *Marine Biology*, **109**, 129–33.

Pérez, M., Duarte, C.M., Romero, J. Sand-Jensen, K. & Alcoverro, T. (1994). Growth plasticity in *Cymodocea nodosa* stands: the importance of nutrient supply. *Aquatic Botany*, **47**, 249–64.

Pérez-Lloréns, J.L. & Niell, F.X. (1995). Short-term phosphate uptake kinetics in *Zostera noltii* Hornem: a comparison between excised leaves and sediment-rooted plants. *Hydrobiologia*, **297**, 17–27.

Pirc, H. (1986). Seasonal aspects of photosynthesis in *Posidonia oceanica*: influence of depth, temperature and light intensity. *Aquatic Botany*, **26**, 203–12.

Ralph, P.J., Gademann, R. & Dennison, W.C. (1998). In situ seagrass photosynthesis measured using a submersible, pulse-amplitude modulated fluorometer. *Marine Biology*, **132**, 367–73.

Redfield, A.C., Ketchum, B.A. & Richards, F.A. (1963). The influence of organisms on the composition of seawater. In *The Sea*, ed. M.N.Hill, Vol.2, pp. 26–77. New York: Wiley.

Sand-Jensen, K., Revsbech, N.P. & Jørgensen, B.B. (1985). Microprofiles of O_2 in epiphyte communities on submerged macrophytes. *Marine Biology*, **89**, 55–62.

Short, F.T., Davis, M.W., Gibson, R.A. & Zimmerman, C.F. (1985). Evidence for phosphorus limitation in carbonate sediments of the seagrass *Syringodium filiforme*. *Estuarine, Coastal and Shelf Science*, **20**, 419–30.

Short, F.T., Dennison, W.C. & Capone, D.G. (1990). Phosphorus-limited growth of the tropical seagrass *Syringodium filiforme* in carbonate sediments. *Marine Ecology Progress Series*, **62**, 169–74.

Short, F.T., Montgomery, J., Zimmerman, C.F. & Short, C.A. (1993). Production and nutrient dynamics of a *Syringodium filiforme* Kütz. seagrass bed in Indian River lagoon, Florida. *Estuaries*, **16**, 323–34.

Smith, R.D., Pregnall, A.M. & Alberte, R.S. (1988). Effects of anaerobiosis on root metabolism of *Zostera marina* (eelgrass): implications for survival in reducing sediments. *Marine Biology*, **98**, 131–41.

Stapel, J. (1997). Nutrient dynamics in Indonesian seagrass beds: factors determining conservation and loss of nitrogen and phosphorus. Ph.D. Thesis, Catholic University, Nijmegen, Netherlands.

Stapel, J. & Hemminga, M.A. (1997). Nutrient resorption from seagrass leaves. *Marine Biology*, **128**, 197–206.

Stapel, J., Aarts, T.L., Van Duynhoven, B.H.M., De Groot, J.D., Van den Hoogen, P.H.W. & Hemminga, M.A. (1996). Nutrient uptake by leaves and roots of the seagrass *Thalassia hemprichii* in the Spermonde Archipelago, Indonesia. *Marine Ecology Progress Series*, **134**, 195–206.

Stapel, J., Manuntun, R. & Hemminga, M.A. (1997). Biomass loss and nutrient redistribution in an Indonesian *Thalassia hemprichii* seagrass bed following seasonal low tide exposure during daylight. *Marine Ecology Progress Series*, **148**, 251–62.

Stumm, W. & Morgan, J.J. (1981). *Aquatic Chemistry. An introduction emphasizing chemical equilibria in natural waters.* 2nd. edn. New York: Wiley.

Terrados, J. & Ros, J.D. (1995). Temperature effects on photosynthesis and depth distribution of the seagrass *Cymodocea nodosa* (Ucria) Ascherson in a Mediterranean coastal lagoon: the Mar Menor (SE Spain). *P.S.Z.N. I: Marine Ecology*, **16**, 133–44.

Terrados, J. & Williams, S.L. (1997). Leaf versus root nitrogen uptake by the surfgrass *Phyllospadix torreyi*. *Marine Ecology Progress Series*, **149**, 267–77.

Terrados, J., Duarte, C.M. & Kenworthy, W.J. (1997). Is the apical growth of *Cymodocea nodosa* dependent on clonal integration? *Marine Ecology Progress Series*, **158**, 103–10.

Thursby, G.B. & Harlin, M.M. (1982). Leaf–root interaction in the uptake of ammonia by *Zostera marina*. *Marine Biology*, **72**, 109–12.

Tomasko, D.A. & Dawes, C.J. (1989). Evidence for physiological integration between shaded and unshaded short shoots of *Thalassia testudinum*. *Marine Ecology Progress Series*, **54**, 299–305.

Udy, J.W. & Dennison, W.C. (1997). Growth and physiological responses of three seagrass species to elevated sediment nutrients in Moreton Bay, Australia. *Journal of Experimental Marine Biology and Ecology*, **217**, 253–77.

Van Lent, F. & Verschuure, J.M. (1994). Intraspecific variability of *Zostera marina* L. (eelgrass) in the estuaries and lagoons of the southwestern Netherlands. I. Population dynamics. *Aquatic Botany*, **48**, 31–58.

Vermaat, J.E. & Verhagen, F.C.A. (1996). Seasonal variation in the intertidal seagrass *Zostera noltii* Hornem.: coupling demographic and physiological patterns. *Aquatic Botany*, **52**, 259–81.

Vermaat, J.E., Beijer, J.A.J., Gijlstra, R., Hootsmans, M.J.M., Phillipart, C.J.M., Van den Brink, N.W. & Van Vierssen, W. (1993). Leaf dynamics and standing stocks of intertidal *Zostera noltii* Hornem. and *Cymodocea nodosa* (Ucria) Ascherson on the Banc d'Arguin. *Hydrobiologia*, **258**, 59–72.

Vermaat, J.E., Agawin, N.S.R., Fortes, M.D., Duarte, C.M., Marbà, N., Enríquez, S. & Van Vierssen, W. (1997). The capacity of seagrasses to survive increased turbidity and siltation: the significance of growth form and light use. *Ambio*, **26**, 499–504.

Wear, D.J., Sullivan, M.J., Moore, A.D. & Millie, D.F. (1999). Effects of water-column enrichment on the production dynamics of three seagrass species and their epiphytic algae. *Marine Ecology Progress Series*, **179**, 201–13.

Zimmerman, R.C. & Alberte, R.S. (1996). Effect of light/dark transition on carbon translocation in eelgrass *Zostera marina* seedlings. *Marine Ecology Progress Series*, **136**, 305–9.

Zimmerman, R.C., Reguzzoni, J.L., Wyllie-Echeverria, S. & Alberte, R.S. (1991). Assessment of environmental suitability for growth of *Zostera marina* L. (eelgrass) in San Francisco bay. *Aquatic Botany*, **39**, 353–66.

Zimmerman, R.C., Reguzzoni, J.L. & Alberte, R.S. (1995). Eelgrass (*Zostera marina* L.) transplants in San Francisco Bay: role of light availability on metabolism, growth and survival. *Aquatic Botany*, **51**, 67–86.

Zimmerman, R.C., Kohrs, D.G., Steller, D.L. & Alberte, R.S. (1997). Impacts of CO_2 enrichment on productivity and light requirements of eelgrass. *Plant Physiology*, **115**, 599–607.

5

Elemental dynamics in seagrass systems

5.1 Introduction

In the previous chapter we discussed the ways seagrasses obtained carbon, nitrogen and phosphorus from the environment, elements that are vital for their structure and functioning. As tissues die, these elements are again lost from the plants, although resorption processes may somewhat mitigate the loss rates. The plants thus have a direct influence on the dynamics of chemical elements in their environment. Uptake by, and loss of, elements from the living plants are only two aspects of the fluxes of matter in seagrass systems. In this chapter we will focus on the various processes determining these fluxes, with particular attention to those relevant to the dynamics of carbon, nitrogen and phosphorus. A variety of processes, biological, physical and chemical, plays a role in shaping the dynamics of these elements, but they share one feature in common: directly or indirectly they are influenced or even determined by the presence of the key species in the system, the seagrasses. Primary production and mineralization are two major processes driving the carbon and nutrient dynamics within the seagrass system, these processes coinciding with fixation and release of inorganic compounds, respectively. Inorganic nitrogen- and phosphorus-containing compounds released during mineralization can be captured again for the production of plant biomass. Although much of the plant biomass dies without being eaten by herbivores and is directly processed by the decomposer community, some of it is consumed by herbivores. This implies a different pathway of metabolic transformations for the elements fixed in plant biomass, but through respiration, excretion and decay processes, these elements are ultimately reintroduced as mineral compounds in the environment as well.

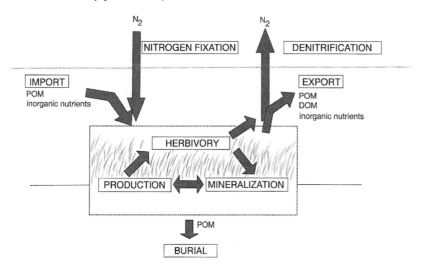

Fig. 5.1. Major processes involved in the elemental dynamics of seagrass systems.

It must be emphasized that the fluxes of matter in seagrass meadows cannot be separated from the environment of these meadows, as if they were processes taking place within a closed system. Seagrass meadows are very much open systems, being components of larger coastal systems and interacting with these environments. Dissolved and particulate matter can enter or leave the system by advective transport in the water, leading to losses or gains of carbon and nutrients. Particulate organic matter may also be buried in the sediment. Atmospheric dinitrogen gas, after dissolution in the water, may be incorporated in organic matter by nitrogen-fixing microflora, a process that results in nitrogen enrichment of the meadows. Reduction of nitrate to dinitrogen gas in the denitrification process, in contrast, may lead to losses of nitrogen from the system. The fluxes of matter that exist between the seagrass system and its environment are therefore important components of the overall pattern of carbon and nutrient dynamics in seagrass systems (Fig. 5.1).

5.2 Primary productivity

The available data on seagrass productivity have recently been compiled and reviewed (Duarte & Chiscano, 1999), and are summarized in Fig. 1.10. In this section only the elemental budgets involved will be briefly outlined, in order to provide a frame of reference for the other fluxes of matter discussed in this chapter.

Aboveground and belowground productivities of seagrasses are significantly correlated. Average aboveground productivities show significant differences between species, ranging between a low value of 0.03 g DW m^{-2} d^{-1} in *Halophila ovalis* to a high value of 14.2 g DW m^{-2} d^{-1} in *Phyllospadix torreyi*. Belowground productivities (roots and rhizomes) also vary among species, ranging from 0.01 g DW m^{-2} d^{-1} in *Halophila ovalis* to 11.3 g DW m^{-2} d^{-1} in *Phyllospadix torreyi*. All data taken together, daily seagrass production averages 3.84 and 1.21 g DW m^{-2} d^{-1} for the aboveground and belowground parts, respectively. The relatively low production value for the belowground parts of the plants probably is an underestimate, as root production is not always included in estimates of belowground productivity. Root production, however, can sometimes be very high, up to nearly half of total plant production (Duarte *et al.*, 1998). Taking the means of the leaf carbon, nitrogen and phosphorus concentrations (33.6, 1.92 and 0.23% DW, respectively; Duarte, 1990), the average nitrogen and phosphorus resorption from seagrass leaves (20.4% and 21.9%, respectively; Hemminga *et al.*, 1999), and assuming a zero resorption of carbon (Stapel & Hemminga, 1997), it can be calculated that the average daily requirements for aboveground production of seagrasses are 1.29 g C, 0.06 g N, and 0.007 g *P* m^{-2} d^{-1}. The requirements for production of roots and rhizomes generally will be lower due to their usually lower areal productivity, and the lower nutrient concentrations in these tissues. It is rather precarious to translate figures on daily requirements to annual requirements, in view of the sometimes profound seasonal fluctuations in productivity (particularly in temperate regions). Yet as a very general approximation of the resource fluxes involved in primary production of seagrass vegetation, it can be stated that the annual carbon uptake typically is in the range of hundreds of grams m^{-2}, the annual nitrogen uptake in the range of tens of grams m^{-2}, and the annual phosphorus uptake in the range of a few grams m^{-2}.

The primary production of seagrasses is a main component of the total primary production in seagrass systems, but it is not the only component. The aboveground parts of the plants are usually colonized by a variety of epiphytic organisms (Fig. 5.2), among which many species of algae, varying in size from microscopically small unicellular forms to macroalgae with centimetres-long thalli. The leaves and stems of the seagrasses offer these algae a suitable substrate for attachment and growth. Being

Fig. 5.2. Heavily epiphytized leaves of *Posidonia oceanica* in the Mediterranean. (Photograph by J. Terrados.)

part of a living organism, however, individual leaves and stems can only act as temporary substrates for the epiphytes. Species diversity of the algae and their biomass therefore may show an increase with life span and age of different plant parts (Borowitzka & Lethbridge, 1989). On individual leaves, a gradient in epiphyte cover is often evident, the older (apical) parts supporting the highest (and most diverse) load. Leaf morphology also plays a role in epiphyte distribution, as is demonstrated by the observation that in co-occurring stands of *Posidonia australis* and *P. sinuosa* the flat leaf surfaces of the former species carry similar epihyte communities on both sides, whereas in the latter species, the concave side of the curved leaf supports a more diverse epiphytic community than the convex side (Trautman & Borowitzka, 1999). The species composition and relative abundance of epiphytic algae, furthermore, depends on environmental conditions, such as exposure to ocean swell, currents and salinity, and also shows seasonal variation (Kendrick *et al.*, 1988; Borowitzka & Lethbridge, 1989; Kendrick & Burt, 1997). Ambient nutrient levels are another critical factor, biomass and productivity of epiphytic algae being increased in nutrient-enriched areas (see section 7.5.1). Epiphytes, when covering the leaf surface, have a detrimental effect on seagrass photosynthesis (Sand-Jensen, 1977). There is no clear evidence, however, that seagrasses are able to control settlement or growth of epiphytes, e.g. by exudation of chemical compounds as is found in various macroalgae. The epiphytes, in contrast, profit from the ample availability of substrate for attachment, the variety of microenvironments in the seagrass canopy, and they may also utilize nitrogen- and phosphorus-containing compounds that are leaking from the leaf surfaces (see section 5.6.1).

Next to epiphytic algae, benthic macroalgae may constitute a diverse part of the canopy of seagrass meadows. Heijs (1985a), for instance, listed as much as 67 benthic macroalgal species accompanying *Thalassia hemprichii* vegetation on a reef flat in Papua New Guinea. The upper layer of the sediment may also harbour a thriving community of benthic microalgae such as diatoms and blue-green algae. Finally, the phytoplankton suspended in the water column adds to total system production. Measurements of the primary productivity of these different components of the seagrass ecosystem are scarce, never comprise all individual autotrophic groups, and have been made in various ways. The available evidence, nonetheless, demonstrates unequivocally that autotrophs other than seagrasses can make a substantial contribution to the total primary production of seagrass meadows (Table 5.1). Epiphyte productivity, the parameter determined most frequently, is typically 20–60% of seagrass aboveground productivity. In some cases the primary production of the seagrasses is even surpassed by that of other components of the autotrophic community (Morgan & Kitting, 1984; Moncreiff *et al.*, 1992; Pollard & Kogure, 1993a).

5.3 Herbivory

The organic matter produced by seagrasses and other primary producers in the seagrass meadow may be utilized by animals living in the meadows or by non-resident species in different ways. Herbivorous animals consume live seagrass blades, epiphytes and macroalgae; filter feeders

Table 5.1. *Comparisons between primary production of seagrasses and other autotrophs in seagrass meadows*

Meadow type	Units	Seagrass	Epiphytes	Benthic microalgae	Phyto-plankton
Cymodocea rotundata[1]	g DW m⁻² d⁻¹	1.28	0.79	–	–
Cymodocea serrulata[1]	g DW m⁻² d⁻¹	2.67	1.69	–	–
Enhalus acoroides[2]	g DW m⁻² d⁻¹	1.75	0.038	–	–
Halodule uninervis[1]	g DW m⁻² d⁻¹	2.66	2.12	–	–
Halodule wrightii[3]	mg C m⁻² h⁻¹	180	229	339	468
Halodule wrightii[4]	g C m⁻² yr⁻¹	256	905	15	7.22
Halodule wrightii[11]	mg C m⁻² h⁻¹	22	0.72	–	–
Posidonia australis[9]	g DW m⁻² yr⁻¹	950–1150	160–211	–	–
Posidonia sinuosa[9]	g DW m⁻² yr⁻¹	700–1000	50–401	–	–
Syringodium isoetif.[1]	g DW m⁻² d⁻¹	3.84	2.07	–	–
Syringodium isoetif.[10]	g C m⁻² d⁻¹	2.2	11.5	4.2	–
Thalassia hemprichit[5]	g DW m⁻² d⁻¹	2.78	1.64	–	–
Thalassia hemprichit[5]	g DW m⁻² d⁻¹	1.61	0.38	–	–
Zostera marina[6]	g C m⁻² d⁻¹	0.9	0.2	–	–
Zostera marina[7]	g C m⁻² yr⁻¹	805	70	–	–
Zostera marina[8]	g C m⁻² yr⁻¹	452 (including epiphytes)		87	54
Zostera marina[12]	g C m⁻² yr⁻¹	707	434	–	–

Seagrass production pertains to aboveground production only.

Data from: (1) Heijs, 1985b; (2) Brouns & Heijs, 1986; (3) Morgan & Kitting, 1984; (4) Moncreiff *et al.*, 1992; (5) Heijs, 1984; (6) Penhale, 1977; (7) Borum *et al.*, 1984; (8) Murray & Wetzel, 1987; (9) Cambridge & Hocking, 1997; (10) Pollard & Kogure, 1993a; (11) Heffernan & Gibson, 1983; (12) Nelson & Waaland, 1997.

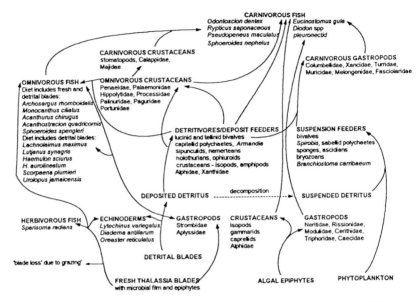

Fig. 5.3. Food web in a *Thalassia testudinum* meadow at Kingston Harbour, Jamaica. (Greenway, 1995.)

capture phytoplankton suspended in the water column; detritivores, finally, feed on dead plant material. The transfer of organic matter from the primary producers to the next trophic level does not stop there, because a variety of carnivorous animals feed upon the primary consumers; in their turn, these may fall prey themselves to other predators. An example of the food web that is the result of these complex feeding relations in a seagrass meadow is shown in Fig. 5.3, based upon the work of Greenway (1995) in a Jamaican *Thalassia testudinum* meadow. Several aspects are noteworthy: first, the large number of species participating in the food web, which clearly illustrates the diverse faunal community that may be encountered in seagrass meadows. In this case the total number of macrobenthic species comprised 20 polychaete species, 39 crustacean species, 61 mollusc species and 40 species belonging to other taxa. In addition, 41 fish species were recorded. We will return to the diversity in seagrass systems in section 6.2. In the second place, it can be seen that the various ways of transfer of organic matter from primary producers to higher trophic levels, as mentioned above, all simultaneously occur in the same meadow. Finally, it is clear that exclusive feeding on living seagrass leaves is a rare feature: in this particular meadow it is restricted to one species only, the fish *Sparisoma radians*. Other species that feed on fresh leaves also depend on other food sources, such as shrimps and crabs in the case of the omnivorous fishes, and detrital plant material in the case of some echinoderms. Gastropods may consume the epidermal layer of fresh seagrass leaves as part of their diet, but they primarily feed on epiphytes growing on the leaves.

The most important common grazers of living seagrass leaves are sea urchins, fish, and water fowl. Other well-known grazers are green turtles (*Chelonia mydas*), the only reptile that consumes substantial quantities of seagrass, and representatives of the Order Sirenia (manatees and dugongs, or sea cows), the only aquatic mammals that feed on seagrass (see section 6.6). The dwindling numbers of these large marine herbivores have put them in the forefront of marine conservation issues, but at the same time have decreased their present-day importance as primary consumers in seagrass systems.

Seagrass leaf consumption by the major invertebrate grazers, sea urchins, has been well documented in tropical and subtropical areas. Herbivory by sea urchins occasionally can be so intense that much of the primary production of the seagrasses is consumed, and may even result in the elimination of extensive seagrass patches (Larkum & West, 1990). The grazing pressure of the urchins is, as can be expected, a function of their population density. A study of the sea urchin *Lytechinus variegatus* in the Gulf of Mexico showed that, depending on the season, 20–40 individuals m^{-2} were required for complete defoliation of *Thalassia testudinum* beds (Valentine & Heck, 1991). As densities of *L. variegatus* in this range or even higher are regularly observed in seagrass beds, it follows that overgrazing events caused by this sea urchin may be quite common. Interestingly, in bare areas predation on young urchins is higher than in vegetated patches, which is expected to result in a decline of the urchin population, and a subsequent recovery of the seagrass vegetation; this, in its turn, may again enhance survival rates of the urchins and lead to population growth (Heck & Valentine, 1995). The grazing pressure exerted by the urchins consequently may show an oscillating pattern in time, due to consecutive cycles of growth and decline of the urchin population. Fluctuating grazing levels are also evident in a study of Philippine seagrass beds. In these beds, the urchins *Tripneustes gratilla* and *Salmacis sphaeroides* are the dominant consumers of live seagrass leaves. *T. gratilla*, in particular, has a clear preference for fresh seagrass leaves (especially *Thalassia hemprichii* leaves), whereas *S. sphaeroides* also consumes considerable amounts of algae and dead seagrass debris. Observations on the population density of *T. gratilla* showed considerable seasonal variation, with densities at peak levels of ca. 9 urchins m^{-2} between November and January, declining to ca. 0.1 m^{-2} between March and July. Using these data, size–ingestion rate relationships, and data on seagrass leaf production, it can be calculated that grazing by *T. gratilla* amounts to 50–100% of seagrass leaf production between November and January, and declines to 1–5% between March and July (Klumpp *et al.*, 1993).

Numerous fish species consume some living and dead seagrass material (McRoy & Helfferich, 1980), but as is exemplified in Fig. 5.3, it should be emphasized that only a few rely primarily on seagrass as a food source. Important consumers of living seagrass leaves can be found among representatives of the Scaridae (parrotfishes), Sparidae (porgies), Monacanthidae (leatherjackets) and Siganidae (rabbitfishes). Examples are *Sparisoma radians* (Scaridae), a small permanent resident of Caribbean seagrass beds; *Lagodon rhomboides* (Sparidae), an abundant species in temperate zone and subtropical seagrass beds along the Atlantic and Gulf coasts of North America (Montgomery & Targett, 1992); the monocanthid *Monacanthus chinensis*, a common herbivore in Australian *Posidonia australia* beds (Conacher *et al.*, 1979), and *Siganus javus*, a siganid from the Indo-Pacific (Pinto & Punchihewa, 1996). Compared with sea urchins, the grazing pressure exerted by fish is less well quantified, but the proportion of seagrass leaf production that is consumed by herbivorous fish generally appears to be low (Klumpp *et al.*, 1993; Greenway,1995; Cebrián *et al.*, 1996) to modest (15%; Havelange *et al.*, 1997). More intense grazing pressures, however, are likely to occur under specific conditions. In field experiments where urchins were excluded from mixed vegetation with *Thalassia testudinum* and the filamentous macroalga *Spyridea hypnoides* as dominant species, leaf consumption by herbivorous fishes (primarily the parrotfish *Sparisoma radians*) equalled production of the seagrass (McGlathery, 1995). These results are consistent with observations in Kenyan coral reef lagoons, which showed that experimental reduction of sea urchin density in protected (unfished) lagoons increased consumption of *Thalassia hemprichii* by parrotfish (McClanahan *et al.*, 1994). Apparently, the extent of grazing by parrotfish is enhanced when sea urchins are absent or present in low densities.

Waterfowl can be major seagrass consumers. Needless to say, consumption by foraging birds is restricted to intertidal or shallow subtidal meadows only. It is, furthermore, often a seasonal phenomenon, coinciding with autumn or winter migrations. An example is the exploitation of intertidal *Zostera marina* and *Zostera noltii* meadows in mid-west European coastal areas by brent goose (*Branta bernicla*), pintail (*Anas acuta*), wigeon (*Anas penelope*) and mallard (*Anas platyrhynchos*), birds that use these areas as a wintering area or as a wintering stop during their migrations to and from arctic breeding grounds in Scandinavia and Russia. The invasion of the *Zostera* beds by the birds may lead to an almost total disappearance of the aboveground biomass (Jacobs *et al.*, 1981), but also to reduction of the belowground biomass, as rhizomes are consumed as well (Tubbs & Tubbs, 1983; Fox, 1996). Another example of the grazing impact of migratory birds is provided by redheads (*Aythya americana*) wintering on the Laguna Madre (Texas). These birds feed almost exclusively on *Halodule wrightii* rhizomes, and consume 75% of the rhizome biomass in shallow (< 0.8 m) meadows (Mitchell *et al.*, 1994).

Grazing on living seagrass plants evidently is highly variable in time and place. Consumption rates may vary between negligible and rates that, at times, more than equal production, leading to complete defoliation of a vegetation. The abundance and composition of the herbivore community obviously are important factors in determining grazing pressure. Moreover, variability is also induced by the fact that not all seagrass species are equally appreciated: herbivores prefer the leaves of species with a relatively high specific growth rate, i.e. species with a high daily leaf production to leaf biomass quotient (Cebrián & Duarte, 1998). The leaves of these species apparently have a higher palatability, perhaps related to lower contents of indigestible lignin-cellulose compounds. Relatively fast growing species from the genera *Cymodocea* and *Halophila*, for example, therefore generally will experience a higher grazing pressure than *Posidonia oceanica* and *Enhalus acoroides*, species which have a very low specific growth rate compared with other seagrasses. Increased leaf nitrogen content may also stimulate consumption by some grazers (McGlathery, 1995). Although the variability in grazing pressure that can be observed in field situations may be high, it is nonetheless likely that the consumption of living leaves represents only a modest pathway in the transformations of organic matter in the majority of seagrass meadows. This can be gathered, for instance, from a compilation of published values on leaf herbivory as percentage of leaf production, which shows that more than half of the recorded values are lower than 15% (Cebrián and Duarte, 1998).

Besides seagrasses, other autotrophs, notably benthic and epiphytic algae, make a significant contribution to total system primary productivity, as we have seen earlier. Much of the carbon in seagrasses is incorporated in cell wall components that are indigestible to most animals. Moreover, physical or chemical maceration mechanisms are necessary to break down the cell walls to be able to assimilate the nutritionally valuable cell contents. Compared with algae, seagrass tissue therefore is a food of poor quality for many animals (see reviews by Thayer *et al.*, 1984, and Klumpp *et al.*, 1989). Consequently, it seems likely that a much higher percentage of the algal production in a seagrass meadow is consumed than generally is the case for the seagrasses. Indeed, scores of animals that primarily use algae can be encountered in seagrass meadows, among them snails, bivalves, isopods, and shrimps (Fig. 5.3). Quantifying the utilization of the algal resources available in a seagrass meadow by a grazing and filter feeding community that may be so diverse obviously is a formidable task. A substantial step forward was made by the detailed studies of Klumpp and coworkers on the grazing of the epiphytic community in tropical, mixed-species seagrass beds (Klumpp *et al.*, 1992, 1993). These authors found that epifaunal grazers (mainly gastropods) consume between 20% and 62% of the net productivity of the algal and bacterial community forming the periphyton on the seagrass leaf surfaces. For reef-flat seagrass meadows, where 23% of the periphyton production was grazed by epifauna and juvenile herbivorous fishes, this implied that in terms of absolute amounts of organic matter consumed, approximately equal amounts of periphyton and living seagrass leaf tissue were consumed. The actual amount of periphyton consumed may even have been higher, because the grazing impact of the abundant shrimps and amphipods in these meadows remained undefined. These data clearly illustrate that grazing food chains in seagrass meadows can be equally or even more dependent on algal productivity than on seagrass productivity.

Part of the plant material that is consumed by herbivores will rapidly be excreted again, after more or less intense metabolic processing. Other constituents will be incorporated in the herbivore's body for a more prolonged period of time. At some stage, however, the nutrients contained in the ingested plant material will return to the environment again, available for plant uptake and growth. This is not necessarily the same site where the herbivore had ingested the original plant material. Excretion processes (and death of the animals) may occur at other places than the foraging site. This is particularly likely for the larger, mobile species, and implies that herbivory probably coincides with some net loss of nutrients from the seagrass beds.

5.4 Benthic mineralization processes

5.4.1 General remarks

Decomposition of aboveground plant tissues occurs partly in the water column. This pertains to the seagrass leaves that start to decompose while still attached to the shoots, but leaf litter lying on top of the sediment can also be counted in this category. For another part, degradation of organic matter occurs in the sediment of the meadow. In this case, the organic matter in question not only comprises dead root and rhizome parts and root exudates, but also leaf fragments and other organic particles deposited on the bottom of the meadow and buried by ongoing sedimentation and bioturbation. Sedimentary inputs, and benthic microalgae and their exudates, are another potentially important source of sedimentary organic matter. Thus, seagrass meadows receive organic inputs of various origins, and this organic matter forms a complicated mix of labile and more refractory compounds. Mineralization of these compounds by heterotrophic organisms involves a chain of reactions in which the organic molecules are stepwise degraded, ultimately to carbon dioxide, the final reaction product. The mineralization of more or less complex organic compounds, e.g. cellulose from seagrass cell wall structures, to carbon dioxide, overall implies the oxidation of carbon, as the carbon in the end product, carbon dioxide, is in a more oxidized form than the carbon in the starting molecule.

Mineralization can only proceed if electron acceptors are available for this oxidation process. In the surface layer of the sediment, dissolved oxygen acts as an electron acceptor. Diffusion of oxygen from the overlying water into the sediment is limited by the slow rate of diffusion through the waterlogged pores, which is approximately four orders of magnitude slower than in gas-filled pores. Dissolved oxygen concentrations thus decrease more or less rapidly with sediment depth, depending on the balance between influx from the sediment surface and removal in the sediment. Increased temperature (which stimulates microbial respiration more than diffusion), high contents of labile organic matter (providing ample substrate for microbial activity), and fine-grained sediment (which restricts the diffusion flux of oxygen) will result in a steep oxygen gradient and a depletion of oxygen already at shallow sediment depths. In coastal marine sediments the oxygenated surface layer often measures just a few centimetres or even millimetres. Deviations of the straightforward pattern of oxygen being restricted to the surface layers of the sediment, however, can be caused by the burrowing and ventilating activities of benthic animals, and by the release of oxygen from plant roots. The result may be a more or less complex mosaic of oxic and anoxic patches in the sediment. Whatever the exact spatial pattern of oxygen availability, aerobic microbial soil processes are replaced by anaerobic processes wherever oxygen becomes limited. The absence of oxygen leads to the use of alternative electron acceptors. A variety of oxidized inorganic compounds can serve this purpose in waterlogged soils. These are not used randomly, but in a sequence that can be predicted on thermodynamic grounds according to the amount of energy that is released as a result of the oxidation of the electron donor. The thermodynamic energy yield suggests, after oxygen depletion, sequential use of NO_3^- or Mn^{4+} (the order of these two oxidants may change depending on conditions), Fe^{3+}, SO_4^{2-}, and CO_2. Whether actual utilization of these electron acceptors in a sediment occurs at all, or in this order, is dependent on such factors as the availability of the various electron acceptors, on competition for electron donors and on inhibiting effects of some oxidants on the reduction of others (Nedwell, 1984). In marine sediments, including in vegetated saltmarsh sediments, the use of SO_4^{2-} (sulphate reduction) is particularly important for anaerobic organic carbon oxidation (Capone & Kiene, 1988; Howarth, 1993).

The total rate of organic matter mineralization in the sediment depends on the contributions of many groups of organisms. In the aerobic zone bacteria, fungi, protozoans and meio- and macrofauna all are capable of completely mineralizing organic molecules to carbon dioxide. Bacteria, which use nitrate as an electron acceptor in the absence of oxygen, are also capable of complete breakdown of complex organic molecules to carbon dioxide. The other bacterial groups that thrive in anaerobic sediment layers are capable of partial oxidation of organic matter only. Complete oxidation of organic macromolecules to carbon dioxide therefore depends on the successive contributions of different parts of the microbial community. The initial step in the breakdown of macromolecules is extracellular enzymatic hydrolysis to smaller units. These units are taken up and metabolized in fermentation reactions to various products, notably short chain fatty acids (particularly acetate), alcohols, H_2 and carbon dioxide. Fermentation does not require the supply of external electron acceptors, unlike the final steps of organic carbon oxidation following fermentation, which can only proceed when coupled to the reduction of electron acceptors derived from the porewater.

In view of the many aspects of benthic mineralization, it is not surprising that the picture of this process in seagrass meadows is far from complete. The following sections are thus necessarily confined to areas where research attention has been focused.

5.4.2 Aerobic and fermentative activity

Application of the thymidine incorporation technique has shed some light on aspects of the bacterial activity in seagrass sediments. With this technique, the incorporation of ^{3}H-labelled thymidine into bacterial DNA is measured, giving information on the division rate of the bacteria, and, hence, on bacterial productivity. The applicability of the technique, however, depends upon the specific bacterial group: division rates of aerobic and fermentative bacteria can be determined with it, but not those of sulphate-reducing bacteria (Gilmore *et al.*,1990; Winding, 1992). Measurements in tropical and subtropical seagrasses show spatial and temporal variation in bacterial productivity. Productivity (of the bacteria for which the procedure works!) is concentrated in the upper 2 cm of the sediment, and decreases with depth (Moriarty *et al.*, 1990; Pollard & Moriarty, 1991; Pollard & Kogure, 1993b). Generally, a clear diurnal pattern in productivity is observed in the upper sediment layer, with peak levels reached during daylight hours. This diurnal pattern disappears at increasing depth in the sediment (Pollard & Moriarty, 1991), and is absent in non-vegetated sediment (Moriarty & Pollard, 1982; Koepfler *et al.*, 1993; Pollard & Kogure, 1993b). The observed fluctuations in bacterial activity are probably determined by the dynamic availability of labile organic compounds. Epibenthic algae in the seagrass bed probably release a part of the carbon fixed during photosynthesis as dissolved organic matter, as has been found similarly in planktonic algae. This organic carbon subsequently can be used by the benthic bacteria. Furthermore, benthic bacterial activity during daylight hours may be stimulated under the influence of solar radiation – increased liberation of dissolved compounds from detrital seagrass leaves lying on the sediment surface (Vähätalo *et al.*, 1998). Finally, when the pattern of a daily increase in bacterial productivity extends beyond the upper 2 cm of the sediment, root exudates probably play a role as well.

Seasonal measurements with the thymidine incorporation technique in the Gulf of Carpentaria (North Australia) showed that productivity levels of bacteria in the sediment surface layers also vary over larger time scales. High levels were established in January/October (summer), and low levels in May/July (winter; Moriarty *et al.*, 1990). This shift in productivity between seasons coincided with a change of 10 °C in mean temperature. Temperature is a major factor controlling bacterial activity, through its impact on the activity of bacterial exoenzymes and cellular metabolism, and may have been the primary cause of seasonal variation, rather than a changing supply of organic matter.

5.4.3 Denitrification

In sediment layers where oxygen becomes scarce or is absent, nitrate in principle is the first alternative electron acceptor. The denitrifying bacteria that are able to rely on nitrate for this purpose are facultative aerobes and they only switch to the use of nitrate in the case of oxygen depletion (Brock & Madigan, 1991). The reduction of nitrate in the denitrification process leads to the production of NO, N_2O, and, in particular, N_2. These gaseous compounds are easily lost from the sediment to the overlying water column. In coastal marine sediments, hence, generally 20–75% of the benthic nitrogen efflux is N_2 (Seitzinger, 1988). Denitrification thus directly affects the nitrogen balance in benthic marine ecosystems as it implies the removal of nitrogen from the system. An adequate supply of nitrate is essential for the denitrification process. The availability of this compound, however, is not obvious in an anoxic sediment environment. Nitrate is supplied in the sediment by three processes: it may diffuse from the water column into the sediment, or enter the sediment by advective transport in groundwater, or it may be produced by nitrification. In the latter process, ammonia is oxidized via nitrite to nitrate by specialized, aerobic bacteria. These nitrifying bacteria are mostly autotrophic; they obtain chemical energy from the oxidations, which enables them to use inorganic carbon as their source of carbon. In most freshwater and coastal marine sediments nitrification probably is the major nitrate-supplying process (Seitzinger, 1988), but discharges of nitrate-polluted groundwater in the coastal zone can be locally important in supplying nitrate to littoral sediments as well (Valiela *et al.*, 1992; Page, 1995).

The prominent role of nitrification in the availability of nitrate in the sediment, and, hence, its importance in supplying substrate for the denitrification process is noteworthy: it implies the coupling of an aerobic process (nitrification) to an anaerobic one (denitrification) in the sediment. Probably, microenvironmental gradients in oxygen are important in the co-occurrence of these two processes in the sediment. Such microgradients may allow simultaneous nitrification and denitrification activities at sediment zones perhaps less than 100 μm apart (Jenkins & Kemp, 1984). The rooted sediment of seagrass beds and other vegetated aquatic systems would seem to be an ideal location for coupled nitrification–denitrification: oxgen leaking from the roots creates oxidized microzones directly adjacent to anoxic sites. Indeed, convincing evidence for enhanced rates of coupled nitrification–denitrification in the sediment of several freshwater and estuarine macrophytes has been presented (Caffrey & Kemp, 1992; Risgaard-Petersen & Jensen, 1997). Yet, the picture that emerges from current investigations is that high rates of coupled nitrification–denitrification are not a consistent feature of aquatic sediments vegetated by rooting angiosperms, but rather are species specific. Coupling between the two processes depends on the extensive release of oxygen by the roots and on the availability of ammonium as a necessary precursor for the production of nitrate by the nitrification process. Oxygen release by the roots, however, differs widely between species, and may be restricted to specific periods of the plant growth cycle. The plants, furthermore, compete with the nitrifying bacteria for ammonium in the sediment, and some may actually deplete the sediment ammonium pool to the disadvantage of the nitrifying bacteria (Verhagen *et al.*, 1995).

Denitrification does occur in seagrass meadows. A recent study on the occurrence of denitrifying bacteria in the sediment of *Thalassia hemprichii* and *Halodule uninervis* showed that these were particularly associated with surfaces of the roots/rhizome complex and the immediate sediment environment (Shieh & Yang, 1997). In a study of nitrification and denitrification in sediment of *Zostera marina* the potential rate of these processes, determined in the laboratory by measuring the process rates in sediment slurries amended with a surplus of substrates (ammonium or nitrate, depending on the process), was generally higher in the seagrass sediment than in adjacent bare sediment. The nitrification and denitrification potentials, however, each showed a distinct seasonal cycle, suggesting that both processes were not strongly coupled in the sediment (Caffrey & Kemp, 1990). *In situ* observations on nitrogen dynamics in eelgrass beds showed that coupled nitrification–denitrification did occur in April, but was not detectable in August. In April it represented only about 25% of the total denitrification rates at that moment (Risgaard-Petersen *et al.*, 1998). Other observations on sediments with the congeneric species *Zostera noltii* likewise demonstrated that coupled nitrification–denitrification rates were not very impressive and only accounted for a minor flow of nitrogen in the sediment (Rysgaard *et al.*, 1996). Thus, until now, there is no strong evidence for a pronounced role of coupled nitrification–denitrification in sediment nitrogen cycling in seagrass meadows. Perhaps a limited release of oxygen by the roots of the seagrass species investigated so far is the reason for this.

Denitrification rates range from less than 1 up to ca. 150 μmol N m^{-2} h^{-1}. These rates are on the low end of the range of denitrification rates measured in other coastal marine systems that commonly fall between 50 and 250 μmol N m^{-2} h^{-1} (Seitzinger, 1988). Denitrification thus generally may only be of modest importance for carbon oxidation in seagrass meadows.

5.4.4 Sulphate reduction

After depletion of nitrate, manganese(IV) and iron(III) oxides may act as electron acceptors for bacterial respiration in anoxic soils (Lovley, 1991). Reduction of these metal oxides may also be a chemical process: sulphide produced during sulphate reduction, and some organic compounds, may directly reduce manganese(IV) and Fe(III) oxides (Laanbroek, 1990). In sediments where these compounds are abundant, chemical reduction may dominate. Iron(III) oxide, for instance, may be reduced by sulphides in salt marsh sediments before it becomes available to bacteria. Hence, its role, and that of manganese(IV) oxide, as terminal electron acceptor is not obvious, and in saltmarsh ecosystems these metal oxides are probably of minor importance (Howarth, 1993). The role of manganese(IV) and Fe(III) oxides in the terminal oxidation reactions of organic matter in seagrass sediments has not been investigated, and can only be a matter of speculation.

The importance of sulphate as electron acceptor is better documented. Sulphate reduction rates in seagrass systems have commonly been determined by measuring the reduction of radioactive ^{35}S-labelled sulphate injected into sediment cores. The rates are derived from the accumulation of ^{35}S in pools of reduced sulphur compounds. In the first place these compounds are H_2S, and HS^-, but reduced ^{35}S may also rapidly react with iron in the sediment, to form iron monosulphide (FeS) or pyrite (FeS$_2$), or end up as elemental sulphur (S$^{\circ}$). The analytical procedures that are applied should account for these compounds as well.

Several features emerge from the studies of sulphate reduction in seagrass meadows. Sulphate reduction rates are much higher in sediments with seagrasses than in bare sediments (Pollard & Moriarty, 1991;

Isaksen & Finster, 1996). Moreover, in a *Zostera marina* meadow, a distinct linear relation between shoot density and sulphate reduction rates was demonstrated (Holmer & Nielsen, 1997). These findings clearly indicate that the presence of the seagrass vegetation directly, or indirectly (e.g. via enhanced sedimentation of organic particles), stimulates sulphate reduction in the sediment. Detailed depth profiles show that sulphate reduction, although an anaerobic process, is already actively proceeding close to the sediment surface, in the upper centimetre. Reduction rates generally attain peak levels in the rooting zone and decrease deeper in the sediment (Fig. 5.4). Although this suggests an involvement of the roots in fuelling the sulphate-reducing bacteria, the details of the root contribution are unclear. Dead root and rhizomes certainly contribute to the larger pool of sediment detritus that is processed by sulphate-reducing bacteria. Some studies, however, indicate that sulphate reduction in the seagrass sediment is stimulated by light, suggesting that organic compounds released by the roots during photosynthesis enhance sulphate reduction (Blackburn *et al.*, 1994; Welsh *et al.*, 1996a; Blaabjerg *et al.*, 1998), but this is not a consistent observation (Isaksen & Finster, 1996). Sulphate reducers can only use a limited array of organic substrates as carbon and energy sources, notably low molecular weight organic acids, alcohols and fatty acids, e.g. lactate, pyruvate, ethanol and acetate. These compounds are generated by fermenting bacteria from organic detritus, but some of them, or their direct precursors, may also be exuded by the roots. Whether a diurnal pattern in sulphate reduction is detectable possibly depends on the amount of suitable substrates for sulphate reducers released by the roots relative to the total production in the sediment. In sediments with a low production, root exudation would be expected to account for a larger proportion of sulphate reduction activity. In this situation, fluctuations in root exudation thus would have a more pronounced and detectable impact on sulphate reduction rates.

At least part of the sulphate-reducing bacterial community is in a spatially favourable position to drain the pool of root exudates. In *Zostera marina* considerable sulphate reduction activity is associated with the roots and rhizomes. This activity virtually disappears when the roots and rhizomes are surface-sterilized with hypochlorite, indicating that most of these sulphate reducers are living on the surface of the seagrass tissues (Blaabjerg & Finster, 1998). This is remarkable in view of the fact that roots of *Z. marina*, like those of other seagrasses, leak oxygen during photosynthesis. It is not clear yet whether the sulphate-

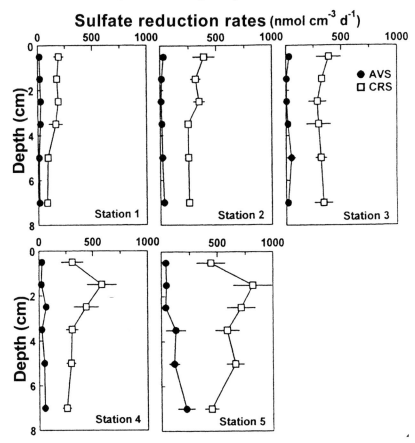

Fig. 5.4. Depth profiles of sulphate reduction rates at 5 stations in Roskilde Fjord, Denmark. Station 1 is an unvegetated site, whereas stations 2–5 are covered by *Zostera marina*, with shoot densities increasing from station 2 to station 5. Sulphate reduction rates are derived from the appearance of ^{35}S in reduced sulphur compounds that are measured as acid-volatile sulphur (AVS) and chromium reducible sulphur (CRS). (Holmer & Nielsen, 1997.)

reducing, root-associated bacteria are confined to root regions where the leakage of oxygen is small or absent. They do show oxygen tolerance, however, in as far as they are able to survive exposure to oxygen, and afterwards regain sulphate reduction rates similar to controls (Blaabjerg & Finster, 1998).

The ability to fix molecular nitrogen (N_2) is a common property among sulphate-reducing bacteria. The sulphate reducers in seagrass sediments are no exception in this respect. In sediments vegetated by

Zostera noltii, addition of molybdate, a specific inhibitor of sulphate-reducing bacteria, led to a more than 80% inhibition of nitrogen fixation (Welsh *et al.*, 1996a). In *Zostera marina* sediments, molybdate was even shown to repress nitrogen fixation by 95%, although in another study a more modest inhibition (25%) was observed (Capone, 1982; McGlathery *et al.*, 1998). To an important extent, sulphate-reducing bacteria and nitrogen-fixing microflora thus may be the same organisms in the sea-grass sediment. If the sulphate-reducing/N_2-fixing bacteria are able to use root exudates, and the seagrass plants can take advantage of the nitrogen that is captured by the bacteria, the co-occurrence of both types of organisms acquires features of a mutualistic relationship, one that is favourable for both. The association of sulphate reducers and plants in the seagrass system, however, is more complicated, because the product of sulphate reduction, sulphide, is a potentially harmful, phytotoxic substance. Investigations in various wetland plants have shown that sulphide may negatively affect aerobic and anaerobic respiratory pro-cesses in the roots, resulting in decreased ATP production (Allam & Hollis, 1972; Koch *et al.*, 1990), and, in the wake of this effect, undoubtedly many other metabolic processes. In *Zostera marina* sulphide levels experimentally elevated above 0.4 mM decreased the potential for utilization of available light (Goodman *et al.*, 1995). Although dissolved sulphide levels are often below 0.1 mM (see section 1.6), such high levels of dissolved sulphide are not completely unrealistic. In the rooting zone of unmanipulated meadows porewater sulphide levels of 0.4–0.7 mM have been measured (Hines, 1991). Even higher levels (2 mM) occur, but such levels coincided with a massive deterioration of the seagrass (*Thalassia testudinum*) vegetation, and may have been the reason for the observed population decline (Carlson *et al.*, 1994). Most often, sulphide levels will be prevented from accumulation to such high levels because sulphide reacts with iron in the sediment, forming insoluble iron-sulphide minerals (FeS and FeS_2), or because the sulphide is microbiologically or chemically oxidized. Carbonate sediments have a very low iron content and a limited capacity to precipitate sulphide (Berner, 1984), and sea-grasses growing on these sediments thus could be particularly prone to sulphide toxicity (Hemminga, 1998). Indeed, the high sulphide levels in the above-mentioned deteriorating *Thalassia testudinum* vegetation (Carlson *et al.*, 1994) were measured in carbonate sediments.

With the data available on denitrification and sulphate reduction rates in seagrass systems, it is possible to obtain an indication of the relative importance of both processes for organic matter oxidation. Under the simplifying assumption that the organic matter that is oxidized has a carbohydrate composition with carbon, hydrogen and oxygen in an atomic ratio of 1:2:1, the equations for denitrification and sulphate reduction are, respectively:

$$5CH_2O + 4NO_3^- \Rightarrow 5CO_2 + 3H_2O + 2N_2 + 4OH^- \tag{1}$$

$$2CH_2O + SO_4^{2-} \Rightarrow 2CO_2 + H_2S + 2OH^- \tag{2}$$

Applying the stoichiometry in these equations, the reduction rates of nitrate and sulphate measured in seagrass sediments can be converted to carbon mineralization rates (Table 5.2). It appears that sulphate reduction is considerably more important than denitrification in this respect. From studies of other, unvegetated, marine sediments it is known that the relative importance of aerobic mineralization and denitrification for carbon oxidation is high at sites with a low carbon input. The importance of sulphate reduction increases with the carbon loading rate of the sediment and a concomitant enhanced total rate of mineralization (Canfield *et al.*, 1993; Wijsman *et al.*, 2000). The dominance of sulphate reduction over denitrification that emerges from the present data, by analogy, may result from a high carbon input to the sediment of the seagrass meadow. The data also suggest that oxygen release from the roots, although it may keep the rhizosphere oxidized (thereby dampening the levels of toxic reduced compounds such as sulphide and iron(II) near the root surfaces), is neither sufficient for advancing the relative importance of denitrification by inducing enhanced rates of coupled nitrification–denitrification in the bulk sediment, nor sufficient in providing enough oxygen to minimize the requirements for sulphate as an alternative electron acceptor.

5.4.5 *The balance between primary production and mineralization*

The coexistence of different primary producers in the seagrass system, and the suite of processes involved in mineralization processes, complicates the question as to whether a balance exists between primary production (carbon production) and mineralization (carbon consumption). However, the number of studies that shed light on this issue is growing. An estimation of total system productivity and mineralization can be obtained by 'simply' measuring the production and consumption of oxygen or carbon dioxide in a meadow. Oxygen is a product of primary production, whereas its consumption by autrophs and heterotrophs

Table 5.2. *Carbon mineralization in seagrass sediments by denitrification and sulphate reduction (integrated* in situ *rates of the rooted sediment layer)*

Seagrass species	Denitrification rate ($mg\ C\ m^{-2}\ d^{-1}$)	Sulphate reduction rate ($mg\ C\ m^{-2}\ d^{-1}$)	Reference
Halodule beaudetti	30–60	300–636	Blackburn *et al.* (1994)
Enhalus acoroides		1800–2160	Pollard & Moriarty (1991)
Zostera marina		312–1644	Blaabjerg *et al.* (1998)
Zostera marina		288–1416	Holmer & Nielsen (1997)
Zostera marina	1.2–12[a]		Iizumi *et al.* (1980)
Zostera marina	1.2– 3.8[b]		Risgaard-Petersen *et al.* (1998)
Zostera novazelandica	2.4–4.8[a]		Kaspar (1983)
Zostera noltii		396–780	Welsh *et al.* (1996a)
Zostera noltii		480	Isaksen & Finster (1996)
Zostera noltii	0.012–4.2[b]		Rysgaard *et al.* (1996)
Syringodium isoetifolium/ Cymodocea serrulata		1464	Pollard & Moriarty (1991)

[a] Recalculated from rates given by Seitzinger (1988).
[b] Denitrification rates in sediment measured after addition of $^{15}NO_3^-$ tracer to overlying water.

provides a quantitative measure of the respiratory processes in a system. This may concern respiratory processes in the oxygenated water column, but also those in the sediment. The latter may seem rather puzzling in view of the use of other electron acceptors than oxygen in anoxic sediment layers. However, when oxygen consumption is determined in order to quantify total benthic respiration, it occurs under the assumption that there is complete reoxidation of the electron acceptors that are reduced in anoxic mineralization processes; this may occur, for instance, when the reduced compounds diffuse to oxic parts of the sediment. Under the assumption of reoxidation, oxygen uptake by the sediment reflects total rates of benthic respiration. Carbon dioxide uptake or production can similarly yield information on primary production and respiration. In this case the proviso is that dissolution or precipitation of carbonate minerals in the sediment and methane production can be neglected.

In a number of studies, transparent benthic chambers have been positioned over seagrass vegetation in order to quantify changes in dissolved oxygen or inorganic carbon due to community metabolism in the meadow (Table 5.3). The data show that in most instances gross primary productivity outweighs respiration. This is apparent from single-measurement studies, but is also demonstrated by studies in which measurements were repeated in different months or seasons. In the few cases that an annual balance could be calculated, productivity exceeded respiration without exception. In other words, seagrass systems usually appear to be net autotrophic: they produce more organic matter in photosynthetic processes than they consume in respiratory processes. Within the same system, a seasonal switch from net autotrophic to net heterotrophic may occur. This is illustrated by the data on the *Posidonia oceanica* meadow presented in the table; in this *Posidonia* meadow the changing balance is due to the fact that plant productivity dominates carbon fluxes in spring, whereas decomposition is maximal later in the year (Frankignoulle & Bouquegneau, 1987).

In systems that are net heterotrophic, mineralization of organic matter outweighs autochthonous production. The relatively high mineralization rates in such systems can only be sustained by imports of allochthonous organic matter. Conversely, the tendency of seagrass systems to be autotrophic suggests that there is generally an internal build-up (burial) of organic carbon, or a net export of organic carbon from these systems, or a combination of both. These processes are discussed in the next sections.

Table 5.3. *Estimates of gross primary production (GPP) and respiration (R) in seagrass systems.* Derived from studies with benthic incubation chambers, supplemented in some cases with measurements of planktonic metabolism in the overlying water column. Results apply to short-term (daily) balances, but if seasonal repetition of the measurements allowed such a calculation, annual balances are also shown.

Species	GPP (g C m^{-2} day^{-1})	R (g C m^{-2} day^{-1})	P–R (g C m^{-2} day^{-1})	P–R (g C m^{-2} yr^{-1})	Source
Zostera marina (April)	0.4	0.4	0		1
Zostera marina (May)	1.5	1.1	0.4		1
Zostera marina (July)	1.8	1.7	0.1		1
Zostera marina (October)	1.2	1.3	−0.1		1
Zostera marina (Winter)	6.6	3.2	3.4		2
Zostera marina (Spring)	3.9	2.3	1.6	} 593	2
Zostera marina (Summer)	4.3	3.2	1.1		2
Zostera marina (Autumn)	2.6	2.1	0.5		2
Posidonia oceanica (Spring)	–	–	0.4		3
Posidonia oceanica (Summer)	–	–	−0.1	} 56	3
Halodule uninervis	3.6	3.0	0.6		4
Thalassia hemprichii	3.1	3.2	−0.1		4
Thalassia/Enhalus	3.1	3.0	0.1		4
Mixed community, Indonesia	3.0	2.9	0.1		5
Mixed community, Indonesia	3.8	3.9	−0.1		5
Mixed community, Indonesia	2.1	2.1	0		5
Mixed community, Indonesia	1.4	2.5	−1.1		5
Mixed community, Sri Lanka	1.1	1.0	0.1		6
Mixed community, Sri Lanka	1.5	1.4	0.1		6
Mixed community, Sri Lanka	0.9	0.2	0.7		6

Table 5.3 (*cont.*)

Species	GPP (g C m^{-2} day^{-1})	R (g C m^{-2} day^{-1})	P–R (g C m^{-2} day^{-1})	P–R (g C m^{-2} yr^{-1})	Source
Thalassia testudinum (June)	2.3	2.2	0.1		7
Thalassia testudinum (Sept.)	2.6	2.6	0	} 40	7
Thalassia testudinum (Jan.)	1.0	0.6	0.4		7
Thalassia testudinum (March)	1.2	0.9	0.3		7

Sources: (1): Lindeboom & de Bree, 1982; measurements included a water column of ca. 30 cm; data read from graphs. (2) Murray & Wetzel, 1987; results are based on benthic and planktonic measurements, and pertain to *Zostera* meadows with an overlying water column of ca. 1 m. (3) Frankignoulle & Bouquegneau, 1987; seasonal data read from graph. (4) Lindeboom & Sandee, 1989; measurements included a water column of ca. 30 cm. (5): Erftemeijer *et al.*, 1993; measurements included a water column of ca. 30 cm. (6): Johnson & Johnstone, 1995; values read from graph. (7): Ziegler & Benner, 1998; results are based on benthic and planktonic measurements, and pertain to *Thalassia* meadows with an overlying water column of ca. 1 m.

5.5 Burial

Mineralization to simple inorganic compounds is ultimately the normal fate of biomass produced by plants, wherever they grow. Part of the organic matter, however, may escape decomposition (at least on relatively short time scales, i.e., up to several years), and is preserved in the sediment. This implies that the carbon and nutrients fixed in the organic constituents are not returned to the ecosystem on biologically relevant time scales, but are essentially lost for ecosystem metabolism. In the biogeochemical literature the long-term storage of locally produced and imported organic matter in sediments is generally referred to as burial. For marine sediments in general, the extent of burial of organic matter appears to be dependent on several factors: the flux of organic matter to the sediment surface; the sediment accumulation rate; the sediment grain size and surface area; the availability of oxygen, and the proportion of refractory substances in the total pool of sediment organic matter (Henrichs, 1993). The fraction of deposited carbon buried in marine sediments may range from less than 1% in some deep-sea sediments to more than 50% at coastal sites (Henrichs, 1993). Much has still to be learned of the actual mechanisms that cause preservation of organic matter in marine sediments. The larger part of the particulate organic matter is strongly associated with the mineral matrix of the sediment (Mayer, 1993). This tight association may be important for preservation, e.g. when organic matter is adsorbed to surfaces in pores smaller than 10 nm, which are inaccessible to microbial exoenzymes (Herman *et al.*, 1999).

Burial of organic matter in seagrass sediments has not received particular attention. The exception is a specific case of burial, i.e. the formation of the peat-like accumulations of coarse seagrass fragments under *Posidonia oceanica* vegetation. Accumulations of persistent organic matter may occasionally develop in sediments of *Thalassodendron ciliatum* and *Posidonia australis* meadows, but truly impressive organic formations are exclusively formed under meadows of the Mediterranean species *Posidonia oceanica*. The horizontal and vertical rhizomes, and the roots of this species, are very decay resistant, as are the leaf sheaths that remain attached to the rhizomes after the leaf blades have died and have been lost from the plant structure. The upward growth of the vertical rhizomes in combination with sedimentation processes leads to the gradual elevation of the seafloor. In this slow process, old and dead roots, rhizomes and attached leaf sheaths become buried deeper and deeper, forming an organic deposit that may be several metres thick and thousands of years old (Boudouresque *et al.*, 1980; Mateo *et al.*, 1997). These deposits (often referred to as 'mattes' in imitation of French authors), thus constitute exceptional long-term records of *Posidonia* meadows and offer many, still largely unexplored, possibilities to gather information on the functioning and persistence of these systems over extended time scales (Hemminga & Mateo, 1996).

Why is the organic debris produced by the deposit-forming species, in particular that of *Posidonia oceanica*, so persistent? Nitrogen and phosphorus contents are known to explain much of the variance in decomposition rates when the different plant groups of the plant kingdom are compared, with decomposition rates increasing with the nitrogen and phosphorus concentration in the tissues of the various plant groups (Enríquez *et al.*, 1993). Observations on the nitrogen and phosphorus concentrations of living rhizomes and 'matte' material demonstrate that these elements are present in rather low concentrations (on a dry weight basis: 0.35–0.86% N and 0.020–0.045% P; Mateo *et al.*, 1997). Furthermore, there is evidence that bacterial activity in *Posidonia oceanica* sediments is nutrient limited (López *et al.*, 1995). Low nutrient availability in the matte thus may well hamper decomposition of the root and rhizome material, but it is unlikely that this explains the long-term persistence of the litter. Probably more important is the rich lignin content of the tissues, in combination with the anoxic conditions of the sediment. Lignin is a polymer build-up mainly from three basic units (monolignols), and it is incorporated in plant cell walls, conferring mechanical rigidity to the plant structure. It is generally thought that the ability to synthesize lignin has been a crucial step in the evolution of land plants. The invasion of the marine environment has not led to the cessation of lignin synthesis in seagrasses. This is not self evident: lignin is a biochemically costly compound, and its presence is not a prerequisite for macrophyte life in the sea, as the absence of lignin in macroalgae demonstrates. Only recently, by application of rigorous analytical protocols, the presence of lignin in seagrasses has been convincingly demonstrated (Opsahl & Benner, 1993; Klap *et al.*, 2000). A comparison among species and tissues showed that lignin abundance varies between species and also between tissues; *Posidonia oceanica* and *Thalassodendron ciliatum* rhizomes contained the highest levels measured (Klap *et al.*, 2000). Lignin is notoriously recalcitrant to microbial degradation, with lignin from some species even more refractory than others. Microbial degradation under anaerobic conditions is possible, but it occurs at much lower rates than under aerobic conditions (Benner *et al.*, 1984). It is therefore likely that high tissue lignin contents and specific lignin properties, in concert with anoxic soil conditions, are the pivotal factors conducive to the gradual accumulation of belowground detritus seen in the deposit-forming species.

Estimations of the rate of accretion of *Posidonia oceanica* deposits, often derived from short-term observations, range widely, from less than a millimetre to more than a centimetre per year. A recent careful study by Mateo *et al.* (1997) in which the age of different layers of *Posidonia* deposits at a number of locations along the Spanish coast were determined by radiocarbon dating, yielded accretion rates between 0.061 and 0.414 cm per year (average: 0.175 cm yr^{-1}). As the observations were all done on deposits with the deepest layers being at least 500 years old, these figures probably give the best approximation of the net annual accretion rates over time scales of centuries and millennia. The long-term avarage accretion rate of 0.175 cm yr^{-1} is equivalent to an annual burial of 58 g C, 0.59 g N and 0.031 g P m^{-2}. These figures are probably conservative estimates because the long-term accretion rate is based on extended periods that may have included erosional events and temporary periods of depressed growth of the meadow. Considering the annual nutrient requirement for biomass production of *Posidonia oceanica* (ca. 7–14 g N m^{-2}, and 0.4–1.2 g P m^{-2}; Romero *et al.*, 1992; Mateo & Romero, 1997), the burial of nitrogen and phosphorus due to matte formation would be equivalent to an annual loss of 4–8% of the nitrogen needed for biomass production, and an annual loss of 3–8% of the required phosphorus.

5.6 Exports and imports of particulate and dissolved matter

5.6.1 *Losses of dissolved organic compounds*

Decomposing seagrass tissues, both live and dead, lose dissolved organic compounds. If these compounds are not intercepted by other organisms in the seagrass system, they are easily exported from the meadow with the flow of the water. Living seagrasses release some of the carbon that is fixed during photosynthesis. The quantities that are released in the water column account for a minor (1–2%) part of the carbon that is incorporated (Penhale & Smith, 1977; Kirchman *et al.*, 1984; Velimirov, 1986; Moriarty *et al.*, 1986). Part of the carbon fixed may also be translocated to the roots and rhizomes and exuded in the sediment. Few data are available on this process, but it may not be quantitatively insignificant:

Moriarty *et al.* (1986) found that within 6 hours 11% of the carbon fixed by leaves of *Halodule wrightii* was exuded into the sediment. Nitrogen and phosphorus are released from the leaves as well. Experiments with ^{32}P tracer indicate that 2–4% of the phosphorus taken up by the root–rhizome system of *Zostera marina* and *Z. noltii* is released through the leaves in the water column (Penhale & Thayer, 1980; Pérez-Lloréns *et al.*, 1993). The epiphytic community directly profits from the various compounds that are leaching from the leaf surface. Using radioactive and stable isotope tracers, it can be demonstrated that carbon, nitrogen and phosphorus move from the plant leaf tissues to algae and bacteria on the leaf surface (Harlin, 1973; McRoy & Goering, 1974; Penhale & Thayer, 1980). Release of dissolved organic carbon (DOC) by *Zostera marina* has been shown to fuel the production of the epiphytic bacterial population (Kirchman *et al.*, 1984). The sediment microbial community similarly will profit from the organic compounds that are exuded from below-ground organs as a corollary of photosynthetic carbon fixation, as is suggested by the diel variability in sediment bacterial activity observed in several studies (see section 5.4).

The limited release of organic compounds from living tissues is followed up by more extensive leaching of organic and inorganic compounds after tissue death. In the context of decomposition, leaching is defined as the abiotic process whereby soluble matter is removed from detritus by the action of water. The release of organic compounds from living tissues and the leaching of dissolved compounds from dead tissues are in principle different processes, yet it is virtually impossible to distinguish their contribution to the total pool of dissolved organic matter in the complex mix of living and dead plant material of the seagrass system. The two processes may even occur simultaneously in the same leaf, as both dead, decomposing, and living tissues can be found in a single leaf (Barnabas, 1992).

Leaching is particularly pronounced during the initial period of decay, and accounts for a large part of the usually considerable weight loss of detritus in this phase. Water-soluble compounds such as monomeric carbohydrates are lost from the cells and easily degradadable constituents of the cell wall structure are solubilized by the action of exoenzymes from bacteria and fungi. Exposure of leaf detritus to solar radiation strongly enhances leaching by degradation of photosensitive leaf components and disintegration of tissue structures (Vähätalo *et al.*, 1998), and particularly in shallow meadows, this may affect the quantities of dissolved compounds that are released. The leaching process, although most prominent during initial decay, will continue through the later decomposition stages. This is partly because the ongoing decay process will gradually expose less accessible tissue parts to microbial attack, and also because the extracellular enzymes of the decomposer community slowly break down more resistant components of the tissue structures as well, generating dissolved compounds.

In the sea the dissolved organic matter released by phytoplankton is an important substrate for growth of water column bacteria; the bacteria are consumed by protozoans which in their turn are preyed upon by zooplankton (Azam *et al.*, 1983). In this way the energy contained in the organic compounds released by the phytoplankton is returned to the main food chain (phytoplankton–zooplankton–fish) via a microbial loop. This chain of events in the water column raises the question of whether dissolved organic compounds leaching from decomposing sea-grass tissues (and from living tissues as well) similarly, via a microbial utilization step, support protozoan grazers and higher trophic levels either in the seagrass system itself, or in the overlying water. Seagrass leaves decomposing in the canopy and on the sediment surface are predominantly colonized by bacteria; the biomass of fungi usually constitutes only a minor component of the total microbial biomass on the litter (Newell, 1981; Blum *et al.*, 1988). As in the open sea, bacteria thus could be the primary connection between seagrass-derived dissolved organic matter and microbial grazers. Organic compounds leaching from seagrass leaves in the early stage of decomposition are readily used and partially converted to bacterial biomass as is shown in experiments with recently detached *Cymodocea nodosa* leaves (Peduzzi & Herndl, 1991). Both free-living bacteria and bacteria attached to the leaf surfaces increased. It was also found that the increased bacterial abundance is followed by development of a rich protozoan community (particularly flagellates), free living in the water of the incubation vessels, and associated with the leaf surface. These observations suggest that a microbial loop starting from the dissolved organic compounds of decaying seagrass leaves is potentially possible. The growth of the biomass of free-living bacteria indicates that the leaf-associated bacteria are not able to utilize all the compounds that are released from the litter. This is in agreement with findings of Blum & Mills (1991). These authors measured bacterial abundance and productivity on decaying leaves of *Zostera marina*, and, simultaneously, the loss rates of carbon from the litter placed in litterbags on the sediment of the seagrass bed. It appeared that detritus-associated bacterial metabolism (the sum of carbon

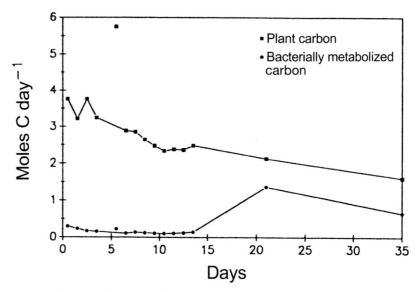

Fig. 5.5. Decay of *Zostera marina* leaves. Relationship between the moles of plant carbon lost per day (filled squares) and the amount of plant carbon metabolized (assimilated and respired) by the bacterial community per day (filled circles). (Blum & Mills, 1991.)

assimilated and carbon respired) accounted for less than 10% of the detrital carbon lost during the first 14 days, rising to about 50% during the later stages of decay (Fig. 5.5). Other data on production of bacteria associated with litter of *Halophila decipiens* indicate that only 0.26% of the daily detrital carbon input in the *Halophila* meadow was converted to bacterial biomass. Even if the bacterial growth efficiency (i.e. the fraction of the metabolized carbon that is allocated to growth) would be a mere 10%, it is clear that bacterial metabolism in this case accounts for only a minor part of litter turn-over (Kenworthy *et al.*, 1989).

It thus seems that the coupling between extracellular enzymatic degradation of litter components by detritus-associated bacteria and processing of these compounds by the same bacteria is not particularly tight: the microbial community in intimate contact with the detritus is only able to take up a modest portion of the plant organic matter; the remaining escapes in the form of dissolved compounds to the water column. This is apparently not only true for the initial phase of decay when a pulse of soluble cellular compounds leaks from the fresh litter, but also for the later, and slower, phases of decay when bacterial exoenzymes continue to liberate organic compounds from the more refractory constituents of the litter. This justifies the conclusion that bacterial production in the water column could play an important role in linking detrital seagrass carbon to other trophic levels. Circumstantial evidence for such a link has been found in a seagrass-dominated lagoon in Texas, where water column respiration was positively correlated with benthic net primary production, the latter process itself being correlated to the net benthic DOC efflux (Ziegler & Benner, 1999), and in some tropical seagrass communities in northern Australia, where bacterial productivities in the water column over the seagrass beds were significantly higher than further offshore where seagrass vegetation was sparse or absent (Moriarty *et al.*, 1990). No differences in bacterioplankton productivity above vegetated and unvegetated sites, however, were detectable in other tropical seagrass meadows (Chin-Leo & Benner, 1991; Pollard & Kogure, 1993b). Furthermore, in Mediterranean *Posidonia oceanica* meadows, planktonic bacterial cell production rates and bacterial abundances within the leaf canopy (where higher dissolved organic matter levels derived from the vegetation are expected) did not differ from those in the water column outside the meadow (Velimirov & Walenta-Simon, 1993; Delille *et al.*, 1996). At present, there are no field data that unambiguously demonstrate a clear link between dissolved organic matter derived from seagrass meadows and bacterial productivity in the water column. It is difficult, of course, to demonstrate such a link in practice, in view of the movement of the water column and the continuous mixing of seagrass-derived dissolved compounds with compounds from other sources, and the potential grazing down of the bacterioplankton by protozoans.

5.6.2 *Export of seagrass litter*

The accumulation of seagrass wrack observed on many beaches presents clear evidence for a common phenomenon: the export of leaves and leaf fragments from seagrass meadows (Fig. 5.6). The deposition of leaf material can be so extensive that the wrack at some places has been collected for commercial purposes: it yields effective insolation material, and can also be processed for the production of soil improvers and fertilizers (Kirkman & Kendrick, 1997). On some beaches deposition of massive amounts of seagrass litter even influences shore geomorphology: the interacting processes of beaching of leaf litter and the trapping and binding of drifting sand may enhance dune formation, as has been found along the Mauritanian coast and in Australia (Hemminga & Nieuwenhuize,

Fig. 5.6. Formation of thick banks of seagrass litter along the Mauritanian coast. (Photograph by M.A. Hemminga.)

1990; Kirkman & Kendrick, 1997). Water currents may transport leafl itter over very large distances, not only landward but also to the open ocean and to great depths. Leaves of *Thalassia testudinum* and other seagrasses originating from meadows in the waters of Florida are carried North by the Gulf Stream and are deposited on the ocean floor off North Carolina at depths of several kilometres (Menzies *et al.*, 1967). The proportion of seagrass leaves that is transported away from a meadow may vary considerably. Estimates determined in different meadows range from less than 1% to almost 80% (Hemminga *et al.*, 1991). Part of this variation can be attributed to differences in leaf characteristics. The cylindrical leaves of *Syringodium filiforme*, for instance, float well and may therefore be easily carried away. In addition, bites of herbivores (e.g. fishes) will completely sever the connection with the parent shoot, instead of merely producing an indentation as usually happens in the case of strap-formed leaves. These factors will contribute to the high leaf export losses (27–79%) observed in this species (Fry & Virnstein, 1988). Waves, water currents, and wind are other factors determining the extent of leaf export, whereas the shoot density of the meadow will also play a role, beds with a high shoot density presumably retaining a larger proportion of the leaves within the canopy. Export of seagrass litter does not always imply a permanent loss of carbon and nutrients from the meadow. The detached leaves accumulating in the surf zone or on the beach are a permanent source of organic particles and dissolved nutrients that are washed back into the sea. Nearshore beds thus may recover at least some of the lost material. If the litter is transported over larger distances, the losses are permanent. Separated from its site of origin, the litter may nonetheless be ecologically important as a valuable food source to benthic deep-sea macrofauna (Wolff, 1979; Suchanek *et al.*, 1985), or in supporting the production of coastal planktonic ecosystems (Thresher *et al.*, 1992). Washed onto the beach, it provides food and shelter to a diverse fauna as well, among them crustaceans, molluscs, insects and polychaetes, which in their turn attract foraging birds (Kirkman & Kendrick, 1992).

5.6.3 Seagrass meadows as traps of particulate matter

It is often observed that the sediment of seagrass meadows is richer in organic matter and in fine-grained silt and clay fractions than is bare sediment outside the beds (e.g. Kenworthy *et al.*, 1982). It is generally assumed that these features are primarily due to the trapping of particles by the canopy (although autochthonous primary production obviously also contributes to the organic enrichment of the sediment). Seagrass canopies change the hydrodynamic conditions of the benthic environment. The moving water in coastal zones loses part of its kinetic energy by interactions with the bottom. The amount of energy that is dissipated depends, among other things, on the roughness of the seafloor. The surface of the seafloor can vary from hydrodynamically smooth, as in the case of fine sands and clays, to very rough. Seagrasses add distinct roughness elements to the benthic surface, and, hence cause dissipation of current energy. Indeed, flume tank studies show that current speed invariably decreases as water enters a patch of seagrass. To give an example, in a *Zostera marina* bed of 15 cm wide and 100 cm long, flow velocity at the downstream end of the bed was only ca. 10–50% of the values measured 10 cm upstream of the bed. Measured below the canopy–water interface, current flux decreased with shoot density. The strongest flow reduction occurred in the first 50 cm of the bed; thereafter little change occurred (Fig. 5.7; Gambi *et al.*, 1990). Observations carried out in natural seagrass beds confirm the occurrence of reduced flow within the vegetation (Worcester, 1995; Gacia *et al.*, 1999). Seagrass meadows also damp vertical wave energy (Fonseca & Calahan, 1992;

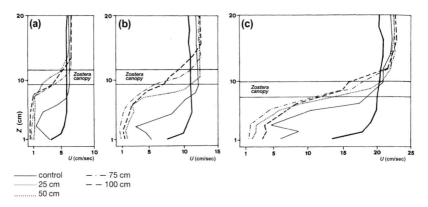

Fig. 5.7. Relationship between flow velocity (U) and height above the bottom (Z) at different distances in *Zostera marina* vegetation in a flume tank. Shoot densities were 1000 m^{-2}. The height of the canopy was approximately 10 cm. Current speeds at the leading edge of the beds were (*a*): 5 cm s^{-1}; (*b*) 10 cm s^{-1}; (*c*): 20 cm s^{-1}. (Gambi *et al.*, 1990.)

Verduin & Backhaus, 2000). The presence of seagrass canopies thus results in hydrodynamically quiet benthic areas. The reduced water motion inside the canopy lowers the particle-carrying capacity of the water, and, consequently, may lead to enhanced particle deposition. In addition, resuspension of sediment particles is expected to be reduced, as this process is particularly evident when current and wave energy is high (Boon *et al.*, 1988; Green *et al.*, 1997). The lowering of sediment resuspension will be further supported by the sediment-binding effect of the dense three-dimensional web formed by the roots and rhizomes.

There are as yet few direct measurements of the particle-trapping effect of natural seagrass meadows. Recently, it was quantified in some Mediterranean *Posidonia oceanica* meadows using sediment traps (Gacia *et al.*, 1999). In this study, a model was applied (Valeur, 1994) that allowed estimation of resuspension rates from measurements of particle flux at different heights in and above the canopy. Fluxes of primary settling matter, i.e. sediment particles deposited for the first time at the bottom of the measuring site, did not differ between meadow and bare sand sites; these varied between ca. 4 and 35 g DW m^{-2} day^{-1}. Fluxes of resuspended sediment, however, were periodically much higher at the bare sand sites that at the meadow site, being up to 47 g DW m^{-2} day^{-1} at the sand site but up to only 3.5 g DW m^{-2} day^{-1} at the meadow sites. These results suggest that the *Posidonia* canopies indeed act as particle traps, but not so much because they enhance deposition, but primarily because they diminish resuspension of particles, in this way reducing sediment erosion and promoting sediment accretion.

Leaf litter export and trapping of particulate matter are not mutually exclusive processes; they can be observed in the same meadow. In a shallow Kenyan bay, where mangroves and seagrasses were present as adjacent systems, it was found that the seagrass beds acted as traps of organic matter outwelling from the mangrove forest during ebb tides, whereas during flood tides seagrass material was taken up in the tidal water flowing over the seagrass area and subsequently entered the mangrove system (Hemminga *et al.*, 1994). The export of seagrass material included leaf macrolitter, and considerable amounts of this material were found on the floor of the mangrove forest (Slim *et al.*, 1996).

Seagrass meadows may not just act as a trap for water column particles because they create a hydrodynamically quiet environment, or because of the belowground network of roots and rhizomes – they may also do so by the filtering activity of associated suspension feeders such as bivalves, ascidians, hydroids, barnacles and sponges (Riisgård *et al.*, 1995; Lemmens *et al.*, 1996). Investigations of *Posidonia sinuosa* meadows in Western Australia suggested that the associated suspension feeding community was capable of filtering the entire 5 m water column over the meadows in about a day. This community thus may well exert a partial control on the load of suspended matter in the water. Meadows of *Amphibolis antarctica* may also be important in this respect, in view of the high densities of epifaunal suspension feeders present on the leaves. Filtration rates in *Heterozostera tasmanica* meadows, however, were expected to be orders of magnitude lower, as suspension feeders only occurred in low densities (Lemmens *et al.*, 1996). The filtering effect of seagrass meadows, as far as it is related to the presence of suspension feeders, thus appears to be variable, and is probably dependent on the type of seagrass vegetation.

5.6.4 *Exchanges of inorganic nutrients*

Undoubtedly, a considerable part of the nutrients contained in, or associated with, organic particles that are captured by suspension feeders, or that are sedimenting in the meadows, will return to the water column as inorganic nutrients, after being liberated as a result of digestion and mineralization processes. Nutrient levels in the sediment porewater of seagrass meadows (derived from allochthonous and from locally produced organic matter) are generally higher than those in the water column (Hemminga, 1998), inducing a net diffusive flux from sediment to water column. Unless the benthic microflora or the seagrass leaves capture these nutrients, they will eventually escape from the system, carried away with the flowing water. This is equally true, of course, for the inorganic nutrients that are released from organic matter decomposing in the water column (e.g. dead leaves that are still attached to the shoots). On the other hand, the seagrasses and other primary producers in the meadow will also intercept external ('new') dissolved inorganic nutrients that are carried with the water flow to the meadow. The capacity of seagrass leaves for nutrient uptake is considerable (see section 4.4.3), and uptake by the leaves can be a dominant element in the nutrient dynamics of a seagrass meadow (Risgaard-Petersen *et al.*, 1998). Nutrient uptake by the leaves is concentration dependent, and the flux of inorganic nutrients from the water column to the vegetation due to leaf uptake is thus most probably less important in nutrient-poor waters than in waters with higher nutrient loads.

5.7 Nitrogen fixation

Seagrass systems may be enriched by the capture of dissolved inorganic compounds by the plants and by the import of particulate matter, but also by a microbial process, nitrogen fixation. The utilization of nitrogen gas (N_2) as a source of nitrogen is a property of only certain bacteria e.g. cyanobacteria. In the fixation process, N_2 is reduced to ammonium and the ammonium subsequently is converted into an organic form. Nitrogen fixation thus results in nitrogen enrichment of an ecosystem. Particularly in nitrogen-limited systems it is therefore potentially important for the overall productivity of the system. The reduction of N_2 is catalyzed by an enzyme complex, nitrogenase, that actually consists of two separate catalytic proteins that reduce nitrogen in successive reduction steps. The process is inhibited by oxygen, yet nitrogen-fixing-microorganisms are found among both the anaerobes and the aerobes. In the latter organisms it is thought that the nitrogenase is active in an O_2-protected microenvironment. This phenomenon is well known in the heterocystous cyanobacteria. These may form specialized cells called heterocysts, which feature thick cell walls that limit the influx of oxygen, and thus allow nitrogenase activity.

Nitrogen fixation is usually demonstrated by the acetylene reduction method. The principle of this method rests on the fact that nitrogenase is capable of reducing acetylene (C_2H_2) to ethylene (C_2H_4), at a rate that is quantitatively analogous to nitrogen reduction (usually it is assumed that the reduction of 3 moles acetylene corresponds to the fixation of 1 mol N_2). The acetylene reduction technique may be used in different ways; for example, an elegant application for measuring nitrogen fixation in sediment that involves minimal disturbance of the sediment system is the brief withdrawal of porewater from sediment cores for acetylene enrichment, followed by reintroduction of the acetylene-amended porewater (McGlathery *et al.*, 1998). All procedures ultimately end in the measurement of the amount of ethylene produced during incubation.

Nitrogen fixation occurs both on the leaves and in the sediment of seagrasses. Heterocystous cyanobacteria probably are primarily responsible for nitrogen fixation on the leaf surface. In agreement with the importance and the largely photoautotrophic character of this group, nitrogen fixation in the phyllosphere usually shows a strong dependence on light and is not enhanced by organic compounds such as glucose. Some observations suggest that apart from cyanobacteria other microbial groups, such as photosynthetic bacteria belonging to the Rhodospirillaceae, may also be involved (Pereg *et al.*, 1994). Nitrogen-fixing bacteria in the sediment of seagrass beds probably form a diverse community. In the upper, oxic, millimetres of the sediment, photoautotrophic heterocystous cyanobacteria and heterotrophic nitrogen-fixing bacteria may co-occur, whereas the deeper, largely anoxic layers of the sediment harbour a mixed species assemblage of heterotrophic nitrogen-fixing bacteria (Pereg *et al.*, 1994). In the root zone of *Zostera marina*, 127 strains of nitrogen-fixing bacteria were isolated, all belonging to the Vibrionaceae (Shieh *et al.*, 1989). The density of these nitrogen-fixing bacteria is particularly high on the roots, and less so in the sediment between the roots, suggesting that the rhizosphere environment provides favourable conditions for the development of a community of bacterial diazotrophs. This is in agreement with observations that nitrogen fixation activity in the rooted sediment of seagrass meadows generally exceeds that of adjacent unvegetated sediment (O'Donohue *et al.*, 1991; Welsh *et al.*, 1996b; McGlathery *et al.*, 1998). In fact, for none of the benthic bacterial processes that we discussed in the previous sections has the close association with the roots and rhizomes become so evident as for nitrogen fixation. A comparison of depth profiles clearly shows the relationship between the root/rhizome biomass and nitrogen fixation activity in an eelgrass sediment, and demonstrates that a seasonal shift of the root/rhizome biomass depth profile coincides with a similar shift in nitrogen fixation activity (Fig. 5.8). Moreover, the figure shows that

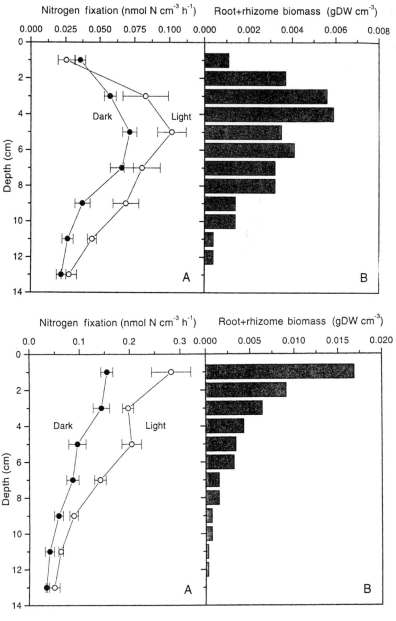

Fig. 5.8. Comparisons of depth profiles of (A) nitrogen fixation in *Zostera marina* vegetated sediment incubated under light and dark conditions and (B) belowground root+rhizome biomass of *Zostera marina* at the spring onset of rapid growth in April (upper panel) and at the summer peak in seagrass biomass in August (lower panel). (McGlathery *et al.*, 1998.)

nitrogen fixation in the sediment is stimulated when the plants are kept in the light, and thus are photosynthetically active. In the same study, glucose was also found to stimulate nitrogen fixation. These observations are in line with results obtained in studies on *Zostera capricorni* and *Zostera noltii* sediments. Nitrogen fixation rates in intact sediment cores with these plants were greater in the light than in the dark, decreased as a result of defoliation (O'Donohue *et al.*, 1991), and were stimulated by sucrose and a range of other carbon compounds (Welsh *et al.*, 1997). The picture that emerges from these data is that the availability of organic substrate is a major factor controlling rates of nitrogen fixation in seagrass sediments, and that the photosynthetically driven exudation of carbon by the plant roots is an important determinant of diel fluctuations in benthic nitrogen fixation rates. The evidence indicating the involvement of seagrass roots in providing carbon compounds to sulphate-reducing bacteria is consistent with this picture, in view of the considerable overlap between nitrogen-fixing bacteria and sulphate-reducing bacteria in the seagrass sediment (see section 5.4.4). The close association of the nitrogen-fixing community with the roots and rhizomes would seem to be problematic in view of the leakage of oxygen from these organs that might inhibit nitrogenase activity. Yet there is no evidence for such an effect. Perhaps spatial separation of oxygen release and nitrogen fixation over microscale distances allows the simultaneous occurrence of both processes, but this is a possibility that needs verification.

Rates of nitrogen fixation in the sediments of seagrass beds are compiled in Table 5.4. Rates vary between 0.1 and 7.3 mg N m^{-2} d^{-1} in temperate meadows and between 7.5 and 47 mg N m^{-2} d^{-1} in tropical and subtropical meadows. Nitrogen fixation thus proceeds at clearly higher rates in subtropical and tropical meadows than in temperate regions. It is not yet possible to give a clear answer to the question if the nitrogen enrichment due to benthic nitrogen fixation in seagrass meadows outweighs the nitrogen losses caused by benthic denitrification. As we have seen earlier, the limited data available on denitrification rates in seagrass sediments yield a range of values up to ca. 150 μmol N m^{-2} h^{-1}. Assuming constant rates for 24 h, this would be equivalent to nitrogen loss rates of up to ca. 50 mg N m^{-2} d^{-1}. In the study that yielded this high denitrification value, nitrogen fixation was also measured and amounted to only 14 mg N m^{-2} d^{-1}, suggesting that nitrogen losses due to denitrification exceeded nitrogen gains due to nitrogen fixation (Blackburn *et al.*, 1994). The denitrification value in this Jamaican

Table 5.4. *Daily rates of nitrogen fixation in the rhizosphere of seagrasses* Nitrogen fixation measured by acetylene reduction.

Species	Site	Nitrogen fixation (mg N m^{-2} d^{-1})	Source
Temperate			
Zostera marina	Limfjord, Denmark	1.2–6.0	McGlathery et al., 1998
Zostera marina	Great South Bay, NY, USA	3.9–6.5	Capone, 1982
Zostera marina	Vaucluse Shores, Virg, USA	5.9	Capone, 1982
Zostera noltii	Bay d'Arcachon, France	0.1–7.3	Welsh et al., 1996a
Tropical / subtropical			
Enhalus acoroides	Gulf of Carpentaria, Australia	25	Moriarty & O'Donohue, 1993
Syringodium isoetifolium/ Cymodocea serrulata	Gulf of Carpentaria, Australia	33	Moriarty & O'Donohue, 1993
Thalassia hemprichii/ Cymodocea rotundata	Gulf of Carpentaria, Australia	16	Moriarty & O'Donohue, 1993
Halodule beaudetti	Oyster Bay, Jamaica	14	Blackburn et al., 1994
Thalassia testudinum	Bimini Harbor, Bahamas	12.1–13.9	Capone & Taylor, 1980
Thalassia testudinum	Biscayne Bay, FL, USA	7.5–24.3	Capone & Taylor, 1980
Zostera capricorni	Moreton Bay, Australia	10–40	O'Donohue et al., 1991

Halodule beaudetti meadow, however, may have been uncommonly high, as the other denitrification values measured in seagrass meadows do not reach higher than 12 mg N m^{-2} d^{-1}. In the only other study where benthic denitrification and nitrogen fixation were simultaneously measured, nitrogen fixation more than compensated for nitrogen losses related to denitrification (Risgaard-Petersen *et al.*, 1998).

Obviously, in balancing the losses and gains of nitrogen, the contributions of nitrogen fixation in the phyllosphere should be considered as well. Data of Moriarty & O'Donohue (1993) in different seagrass communities in the tropical Gulf of Carpentaria show that leaf epiphytes may lead to enrichment of the meadows with 0.3–4.2 mg N m^{-2} d^{-1}, accounting for up to 13% of the nitrogen fixation activity in these seagrass beds. Earlier data present a rather inconsistent picture, with estimations that are widely varying. The quantitative contribution of nitrogen fixation in the phyllosphere to the nitrogen enrichment of seagrass meadows therefore is rather obscure at present.

5.8 Concluding remarks

In this chapter we have discussed the many processes that contribute to elemental dynamics in seagrass meadows. The number of processes is such that no single study can address all of them simultaneously. In addition, the compilation of data on specific process rates shows that these rates differ considerably among meadows, often by more than one order of magnitude. It is thus a precarious matter to define the rates of the individual processes for the 'average' seagrass meadow, and we will not attempt this. There are several studies, however, that give insight into the net result of the various processes. The studies on total system productivity and mineralization indicate that the fixation of carbon in organic matter usually exceeds the mineralization of organic matter in seagrass systems (section 5.4.5). Seagrass meadows thus tend to be net autotrophic, a feature that they share with other productive marine ecosystems (Duarte & Agusti, 1998; Gattuso *et al.*, 1998). Such a pattern is suggestive of net export of organic carbon, or an internal build-up of organic carbon in these systems. Indeed, both export of leaf material and accumulation of refractory plant material have been observed in seagrass meadows (sections 5.5 and 5.6).

However, the studies balancing primary production and mineralization give no *definite* answer to the question of whether seagrass meadows show a net gain of carbon, nitrogen and phosphorus in the course of time. Sedimentation of refractory organic matter, for instance, would gradually increase the organic carbon pool in the meadow without having much influence on mineralization rates. Direct observations on pool sizes are more useful in that respect. The study of Kenworthy *et al.* (1982) already suggested that colonization of bare sediment by *Zostera marina* and *Halodule wrightii* led to a net increase in the pool of sediment organic matter and of total sediment nitrogen. The recent study of Risgaard-Petersen *et al.* (1998) on a Danish *Zostera marina* meadow confirms this for nitrogen. These authors found that, from the beginning of the growing season (April) until the month of maximum productivity (August), this meadow acts as a net sink of nitrogen, with an average daily accumulation rate of 290 mg N m^{-2}. The dominant source of this accumulating nitrogen was suspended matter that sedimented into the bed. The next important process was nitrogen uptake by the leaves. Nitrogen fixation, another nitrogen flux to the bed, accounted for only a minor part of the nitrogen enrichment. The pronounced nitrogen enrichment observed in this meadow probably is only a seasonal phenomenon. Much of the nitrogen may be lost when detached leaves are exported *en masse* and the sediment is eroded by autumn and winter storms. A net nitrogen increase over time scales of years, however, is possible in seagrass meadows. Colonization and development of *Cymodocea nodosa* stands coincided with a yearly net accumulation of 3.6 g N m^{-2} over a period of 5 years (Pedersen *et al.*, 1997). Plant-bound nitrogen alone could not explain this increase. Hence, the presence of *Cymodocea* must have led to the import of nitrogen, presumably mainly via leaf uptake or increased sedimentation.

In several situations it is thus clear that the net effect of the various flux processes operating in seagrass meadows is an accumulation of carbon and nitrogen. With respect to phosphorus, the situation is less clear, particularly because this element has received less attention. The development of the phosphorus pool can be different from that of carbon and nitrogen; physical and chemical processes are relatively more important for the dynamics of phosphorus than for those of carbon and nitrogen, which are strongly determined by (micro)biological processes. In the case of the *Cymodocea nodosa* stands mentioned above, the total phosphorus pool, in contrast to the total carbon and nitrogen pool, was unaffected by plant development (Pedersen *et al.*, 1997).

A major gap in our knowledge on the development of carbon and nutrient pools, furthermore, concerns the situation in oligotrophic, tropical seagrass meadows. The seasonally more stable climate, and the low water column nutrient levels and particle contents, would seem to render some of the processes that feature prominently in the elemental dynamics of temperate and subtropical meadows *a priori* less important. However, the recent data on leaf uptake and on sedimentation in tropical meadows (see sections 4.4.3 and 7.5.2) indicate that surprises can be expected.

5.9 References

Allam, A.I. & Hollis, J.P. (1972). Sulfide inhibition of oxidases in rice roots. *Phytopathology*, **62**, 634–9.

Azam, F., Fenchel, T., Field, J.G., Gray, J.S., Meyer-Reil, L.A. & Thingstad, F. (1983). The ecological role of water-column microbes in the sea. *Marine Ecology Progress Series*, **10**, 257–63.

Barnabas, A.D. (1992). Bacteria on and within leaf blade epidermal cells of the seagrass *Thalassodendron ciliatum* (Forssk.) Den Hartog. *Aquatic Botany*, **43**, 257–66.

Benner, R., Maccubbin, A.E. & Hodson, R.E. (1984). Anaerobic biodegradation of the lignin and polysaccharide components of lignocellulose and synthetic lignin by sediment microflora. *Applied and Environmental Microbiology*, **47**, 998–1004.

Berner, R.A. (1984). Sedimentary pyrite formation: an update. *Geochimica and Cosmochimica Acta*, **48**, 605–15.

Blaabjerg, V. & Finster, K. (1998). Sulphate reduction associated with roots and rhizomes of the marine macrophyte *Zostera marina*. *Aquatic Microbial Ecology*, **15**, 311–14.

Blaabjerg, V., Mouritsen, K.M. & Finster, K. (1998). Diel cycles of sulphate reduction rates in sediments of a *Zostera marina* bed (Denmark). *Marine Ecology Progress Series*, **15**, 97–102.

Blackburn, T.H., Nedwell, D.B. & Wiebe, W.J. (1994). Active mineral cycling in a Jamaican seagrass sediment. *Marine Ecology Progress Series*, **110**, 233–9.

Blum, L.K. & Mills, A.L. (1991). Microbial growth and activity during the initial stages of seagrass decomposition. *Marine Ecology Progress Series*, **70**, 73–82.

Blum, L.K., Mills, A.L., Zieman, J.C. & Zieman, R.T. (1988). Abundance of bacteria and fungi in seagrass and mangrove detritus. *Marine Ecology Progress Series*, **42**, 73–8.

Boon, J.D., Green, M.O. & Suh, K. (1988). Bimodal wave spectra in the lower Chesapeake Bay, sea bed energies and sediment transport during winter storms. *Continental (and) Shelf Research*, **16**, 1965–88.

Borowitzka, M.A. & Lethbridge, R.C. (1989). Seagrass epiphytes. In *Biology of Seagrasses*, ed. A.W.D Larkum, A.J. McComb & S.A. Shepherd, pp. 458–85. Amsterdam: Elsevier.

Borum, J., Kaas, H. & Wium-Andersen, S. (1984). Biomass variation and autotrophic production of an epiphyte–macrophyte community in a coastal Danish area: II. Epiphyte species composition, biomass and production. *Ophelia*, **23**, 165–79.

Boudouresque, C.F., Giraud, G., Thommeret, J. & Thommeret, Y. (1980). First attempt at dating by ^{14}C the undersea beds of dead *Posidonia oceanica* in the bay of Port-Man (Port-Cros, Var, France). *Traveaux scientifiques du Parc National de Port-Cros, France*, **6**, 239–42.

Brock, T.D. & Madigan, M.T. (1991). *Biology of Microorganisms*. Prentice-Hall, London.

Brouns, J.J.W.M. & Heijs, F.M.L. (1986). Production and biomass of the seagrass *Enhalus acoroides* (L.f.) Royle and its epiphytes. *Aquatic Botany*, **25**, 21–45.

Caffrey, J.M. & Kemp, W.M. (1992). Influence of the submersed plant, *Potamogeton perfoliatus*, on nitrogen cycling in estuarine sediments. *Limnology and Oceanography*, **37**, 1483–95.

Cambridge, M.L. & Hocking, P.J. (1997). Annual primary production and nutrient dynamics of the seagrasses *Posidonia sinuosa* and *Posidonia australis* in south-west Australia. *Aquatic Botany*, **59**, 277–95.

Canfield, D.E., Thamdrup, B. & Hansen, J.W. (1993). The anaerobic degradation of organic matter in Danish coastal sediments: iron reduction, manganese reduction, and sulfate reduction. *Geochimica et Cosmochimica Acta*, **57**, 3867–83.

Capone, D.G. (1982). Nitrogen fixation (acetylene reduction) by rhizosphere sediments of the eelgrass *Zostera marina*. *Marine Ecology Progress Series*, **10**, 67–75.

Capone, D.G. & Kiene, R.P. (1988). Comparison of microbial dynamics in marine and freshwater sediments: contrasts in anaerobic carbon metabolism. *Limnology and Oceanography*, **33**, 725–49.

Capone, D.G. & Taylor, B.F. (1980). N_2 fixation in the rhizosphere of *Thalassia testudinum*. *Canadian Journal of Microbiology*, **26**, 998–1005.

Carlson, P.R., Yabro, L.A., Barber, T.R. (1994). Relationship of sediment sulfide to mortality of *Thalassia testudinum* in Florida Bay. *Bulletin of Marine Science*, **54**, 733–46.

Cebrián, J. & Duarte, C.M. (1998). Patterns in leaf herbivory on seagrasses. *Aquatic Botany*, **60**, 67–82.

Cebrián, J., Duarte, C.M., Marbà, N., Enríquez, S., Gallegos, M. & Olesen, B. (1996). Herbivory on *Posidonia oceanica*: magnitude and variability in the Spanish Mediterranean. *Marine Ecology Progress Series*, **130**, 147–55.

Chin-Leo, G. & Benner, R. (1991). Dynamics of bacterioplankton abundance and production in seagrass communities of a hypersaline lagoon. *Marine Ecology Progress Series*, **73**, 219–30.

Conacher, M.J., Lanzing, W.J.R. & Larkum, A.W.D. (1979). Ecology of Botany Bay II. Aspects of the feeding ecology of the fanbellied leatherjacket *Monacanthus chinensis* in *Posidonia australis* beds. *Australian Journal of Marine and Freshwater Research*, **30**, 387–400.

Delille, D., Canon, C. & Windeshausen, F. (1996). Comparison of planktonic and benthic bacterial communities associated with a Mediterranean *Posidonia* seagrass system. *Botanica Marina*, **39**, 239–49.

Duarte, C.M. (1990). Seagrass nutrient content. *Marine Ecology Progress Series*, **67**, 201–7.

Duarte, C.M. & Agusti, S. (1998). The CO_2 balance of unproductive aquatic ecosystems. *Science*, **281**, 234–6.

Duarte, C.M. & Chiscano, C.L. (1999). Seagrass biomass and production: a reassessment. *Aquatic Botany*, **65**, 123–39.

Duarte, C.M., Merino, M., Agawin, N.S.R., Uri, J., Fortes, M.D., Gallegos, M.E., Marbà, N. & Hemminga, M.A. (1998). Root production and belowground biomass. *Marine Ecology Progress Series*, **171**, 97–108.

Enríquez, S., Duarte, C.M. & Sand-Jensen, K. (1993). Patterns in decomposition rates among photosynthetic organisms: the importance of detritus C:N:P content. *Oecologia*, **94**, 457–71.

Erftemeijer, P.L.A., Osinga, R. & Mars, A.E. (1993). Primary production of seagrass beds in South Sulawesi (Indonesia): a comparison of habitats, methods and species. *Aquatic Botany*, **46**, 67–90.

Fonseca, M.S. & Calahan, J.A. (1992). A preliminary evaluation of wave attenuation by four species of seagrasses. *Estuarine, Coastal and Shelf Science*, **35**, 565–76.

Fox, A.D. (1996). *Zostera* exploitation by Brent Geese and Wigeon on the Exe Estuary, southern England. *Bird Study*, **43**, 257–68.

Frankignoulle, M. & Bouquegneau, J.M. (1987). Seasonal variation of the diel carbon budget of a marine macrophyte ecosystem. *Marine Ecology Progress Series*, **38**, 197–9.

Fry, B. & Virnstein, R.W. (1988). Leaf production and export of the seagrass *Syringodium filiforme* Kutz. in Indian River Lagoon, Florida. *Aquatic Botany*, **30**, 261–66.

Gacia, E., Granata, T.C. & Duarte, C.M. (1999). An approach to measurement of particle flux and sediment retention within seagrass (*Posidonia oceanica*) meadows. *Aquatic Botany*, **65**, 255–68.

Gambi, M.C., Nowell, A.R.M. & Jumars, P.A. (1990). Flume observations on flow dynamics in *Zostera marina* (eelgrass) beds. *Marine Ecology Progress Series*, **61**, 159–69.

Gattuso, J.P., Frankignoulle, H. & Wollast, R. (1998). Carbon and carbonate metabolism in coastal aquatic ecosystems. *Annual Review of Ecology and Systematics*, **29**, 405–34.

Gilmore, C.C., Leavit, M.E. & Shiaris, M.P. (1990). Evidence against incorporation of exogenous thymidine by sulfate reducing bacteria. *Limnology and Oceanography*, **34**, 1401–9.

Goodman, J.L., Moore, K.A. & Dennison, W.C. (1995). Photosynthetic responses of eelgrass (*Zostera marina* L.) to light and sediment sulfide in a shallow barrier island lagoon. *Aquatic Botany*, **50**, 37–47.

Green, M.O., Black, K.P. & Amos, C.L. (1997). Control of estuarine sediment dynamics by interactions between currents and waves at several scales. *Marine Geology*, **144**, 97–114.

Greenway, M. (1995). Trophic relationships of macrofauna within a Jamaican seagrass meadow and the role of the echinoid *Lytechinus variegatus* (Lamarck). *Bulletin of Marine Science*, **56**, 719–36.

Harlin, M.M. (1973). Transfer of products between epiphytic marine algae and host plants. *Journal of Phycology*, **9**, 243–8.

Havelange, S., Lepoint, G., Dauby, P. & Bouquegneau, J.-M. (1997). Feeding of the Sparid fish *Sarpa salpa* in a seagrass ecosystem: diet and carbon flux. *P.S.Z.N. I: Marine Ecology*, **18**, 289–97.

Heck, K.L. & Valentine, J.F. (1995). Sea urchin herbivory: evidence for long-lasting effects in subtropical seagrass meadows. *Journal of Experimental Marine Biology and Ecology*, **189**, 205–17.

Heffernan, J.J. & Gibson, R.A. (1983). A comparison of primary production rates in Indian River, Florida seagrass systems. *Florida Science*, **46**, 286–95.

Heijs, F.M.L. (1984). Annual biomass and production of epiphytes in three monospecific seagrass communities of *Thalassia hemprichii* (Ehrenb.) Aschers. *Aquatic Botany*, **20**, 195–218.

Heijs, F.M.L. (1985a). The macroalgal component in monospecific seagrass beds from Papua New Guinea. *Aquatic Botany*, **22**, 291–324.

Heijs, F.M.L. (1985b). Some structural and functional aspects of the epiphytic component of four seagrass species (Cymodoceoideae) from Papua New Guinea. *Aquatic Botany*, **23**, 225–47.

Hemminga, M.A. (1998). The root/rhizome system of seagrasses: an asset and a burden. *Journal of Sea Research*, **39**, 183–196.

Hemminga, M.A. & Mateo, M.A. (1996). Stable carbon isotopes in seagrasses: variability in ratios and use in ecological studies. *Marine Ecology Progress Series*, **140**, 285–98.

Hemminga, M.A. & Nieuwenhuize, J. (1990). Seagrass wrack-induced dune formation on a tropical coast (Banc d'Arguin, Mauritania). *Estuarine, Coastal and Shelf Science*, **31**, 499–502.

Hemminga, M.A., Harrison, P.G. & van Lent, F. (1991). The balance of nutrient losses and gains in seagrass meadows. *Marine Ecology Progress Series*, **71**, 85–96.

Hemminga, M.A., Slim, F.J., Kazungu, J., Ganssen, G.M., Nieuwenhuize, J. & Kruyt, N.M. (1994). Carbon outwelling from a mangrove forest with adjacent seagrass beds and coral reefs (Gazi Bay, Kenya). *Marine Ecology Progress Series*, **106**, 291–301.

Hemminga, M.A., Marbà, N., & Stapel, J. (1999). Leaf nutrient resorption, leaf lifespan and the retention of nutrients in seagrass systems. *Aquatic Botany*, **65**, 141–58.

Henrichs, S.M. (1993). Early diagenesis of organic matter: the dynamics (rates) of cycling of organic compounds. In *Organic Geochemistry*, ed. M.H. Engel & S.A. Macko, pp. 101–17. New York: Plenum Press.

Herman, P.M.J., Middelburg, J.J., Van de Koppel, J. & Heip, C.H.R. (1999). Ecology of estuarine macrobenthos. *Advances in Ecological Research*, **29**, 195–240.

Hines, M.E. (1991). The role of certain infauna and vascular plants in the mediation of redox reactions in marine sediments. In *Developments in Geochemistry*, vol. 6, *Diversity of Environmental Biogeochemistry*, ed. J. Berthelin, pp. 275–86. Amsterdam: Elsevier.

Holmer, M. & Nielsen, S.L. (1997). Sediment sulfur dynamics related to biomass-density patterns in *Zostera marina* (eelgrass) beds. *Marine Ecology Progress Series*, **146**, 163–71.

Howarth, R.W. (1993). Microbial processes in salt-marsh sediments. In *Aquatic Microbiology*, ed. T. Edgcumbe Ford, pp. 239–61. Oxford: Blackwell.

Iizumi, H., Hattori, A. & McRoy, C.P. (1980). Nitrate and nitrite in interstitial waters of eelgrass beds in relation to the rhizosphere. *Journal of Experimental Marine Biology and Ecology*, **47**, 191–201.

Isaksen, M.F. & Finster, K. (1996). Sulphate reduction in the root zone of the seagrass *Zostera noltii* on the intertidal flats of a coastal lagoon (Arcachon, France). *Marine Ecology Progress Series*, **137**, 187–94.

Jacobs, R.P.W.M., Den Hartog, C., Braster, B.F. & Carriere, F.C. (1981). Grazing of the seagrass *Zostera noltii* by birds at Terschelling (Dutch Wadden Sea). *Aquatic Botany*, **10**, 241–59.

Jenkins, M.C. & Kemp, W.M. (1984). The coupling of nitrification and denitrification in two estuarine sediments. *Limnology and Oceanography*, **29**, 609–19.

Johnson, P. & Johnstone, R. (1995). Productivity and nutrient dynamics of tropical sea-grass communities in Puttalam Lagoon, Sri Lanka. *Ambio*, **24**, 411–17.

Kaspar, H.F. (1983). Denitrification, nitrate reduction to ammonium, and inorganic nitrogen pools in intertidal sediments. *Marine Biology*, **74**, 133–9.

Kendrick, G.A. & Burt, J.S. (1997). Seasonal changes in epiphytic macro-algae assemblages between offshore exposed and inshore protected *Posidonia sinuosa* Cambridge *et* Kuo seagrass meadows, Western Australia. *Botanica Marina*, **40**, 77–85.

Kendrick, G.A., Walker, D.I. & McComb, A.J. (1988). Changes in distribution of macro-algal epiphytes on stems of the seagrass *Amphibolis antarctica* along a salinity gradient in Shark Bay, Western Australia. *Phycologia*, **27**, 201–8.

Kenworthy, W.J., Zieman, J.C. & Thayer, G.W. (1982). Evidence for the influence of seagrasses on the benthic nitrogen cycle in a coastal plain estuary near Beaufort, North Carolina (USA). *Oecologia*, **54**, 152–8.

Kenworthy, W.J., Currin, C.A., Fonseca, M.S. & Smith, G. (1989). Production, decomposition, and heterotrophic utilization of the seagrass *Halophila decipiens* in a submarine canyon. *Marine Ecology Progress Series*, **51**, 277–90.

Kirchman, D.L., Mazella, L., Alberte, R.S. & Mitchell, R. (1984). Epiphytic bacterial production on *Zostera marina*. *Marine Ecology Progress Series*, **15**, 117–23.

Kirkman, H. & Kendrick, G.A. (1997). Ecological significance and commercial harvesting of drifting and beach-cast macro-algae and seagrasses in Australia: a review. *Journal of Applied Phycology*, **9**, 311–26.

Klap, V.A., Hemminga, M.A. & Boon, J.J. (2000). The retention of lignin in seagrasses: angiosperms that returned to the sea. *Marine Ecology Progress Series* **194**; 1–11.

Klumpp, D.W., Howard, R.K. & Pollard, P.C. (1989). Trophodynamics and nutritional ecology of seagrass communities. In *Biology of Seagrasses*, ed. A.W.D Larkum, A.J. McComb & S.A. Shepherd, pp. 394–457. Amsterdam: Elsevier.

Klumpp, D.W., Salita-Espinosa, J.T. & Fortes, M.D. (1992). The role of epiphytic periphyton and macroinvertebrate grazers in the trophic flux of a tropical seagrass community. *Aquatic Botany*, **43**, 327–49.

Klumpp, D.W., Salita-Espinosa, J.T. & Fortes, M.D. (1993). Feeding ecology and trophic role of sea urchins in a tropical seagrass community. *Aquatic Botany*, **45**, 205–29.

Koch, M.S., Mendelssohn, I.A., McKee, K.L. (1990). Mechanism for the hydrogen sulfide-induced growth limitation in wetland macrophytes. *Limnology and Oceanography* **35**, 399–408.

Koepfler, E.T., Benner, R. & Montagna, P.A. (1993). Variability of dissolved organic carbon in sediments of a seagrass bed and an unvegetated area within an estuary in Southern Texas. *Estuaries* **16**, 391–404.

Laanbroek, H.J. (1990). Bacterial cycling of minerals that affect plant growth in waterlogged soils: a review. *Aquatic Botany*, **38**: 109–25.

Larkum, A.W.D. & West, R.J. (1990). Long-term changes of seagrass meadows in Botany Bay, Australia. *Aquatic Botany*, **37**, 55–70.

Lemmens, J.W.T.J., Clapin, G., Lavery, P. & Cary, J. (1996). Filtering capacity of seagrass meadows and other habitats of Cockburn Sound, Western Australia. *Marine Ecology Progress Series*, **143**, 187–200.

Lindeboom, H.J. & de Bree, B.H.H. (1982). Daily production and consumption in an eelgrass (*Zostera marina*) community in saline Lake Grevelingen: discrepancies between the O_2 and ^{14}C method. *Netherlands Journal of Sea Research*, **16**, 362–79.

Lindeboom, H.J. & Sandee, A.J.J. (1989). Production and consumption of tropical seagrass fields in Eastern Indonesia measured with bell jars and microelectrodes. *Netherlands Journal of Sea Research*, **23**, 181–90.

López, N.I., Duarte, C.M., Vallespinós, F., Romero, J. & Alcoverro, T. (1995). Bacterial activity in NW Mediterranean seagrass (*Posidonia oceanica*) sediments. *Journal of Experimental Marine Biology and Ecology*, **187**, 39–49.

Lovley, D.R. (1991). Dissimilatory Fe(III) and Mn(IV) reduction. *Microbiological Reviews*, **55**, 259–87.

Mateo, M.A. & Romero, J. (1997). Detritus dynamics in the seagrass *Posidonia oceanica*: elements for an ecosystem carbon and nutrient budget. *Marine Ecology Progress Series*, **151**, 43–53.

Mateo, M.A., Romero, J., Pérez, M., Littler, M.M. & Littler, D.S. (1997). Dynamics of millenary organic deposits resulting from the growth of the Mediterranean seagrass *Posidonia oceanica*. *Estuarine, Coastal and Shelf Science*, **14**, 103–10.

Mayer, L.M. (1993). Organic matter at the sediment–water interface. In *Organic Geochemistry*, ed. M.H. Engel & S.A. Macko, pp. 171–84. New York: Plenum Press.

McClanahan, T.R., Nugues, M. & Mwachireya, S. (1994). Fish and sea urchin herbivory and competition in Kenyan coral reef lagoons: the role of reef management. *Journal of Experimental Marine Biology and Ecology*, **184**, 237–54.

McGlathery, K.J. (1995). Nutrient and grazing influences on a subtropical seagrass community. *Marine Ecology Progress Series*, **122**, 239–52.

McGlathery, K.J., Risgaard-Petersen, N. & Christensen, P.B. (1998). Temporal and spatial variation in nitrogen fixation activity in the eelgrass *Zostera marina* rhizosphere. *Marine Ecology Progress Series*, **168**, 245–58.

McRoy, C.P. & Goering, J.J. (1974). Nutrient transfer between the seagrass *Zostera marina* and its epiphytes. *Nature*, **248**, 173–4.

McRoy, C.P. & Helfferich, C. (1980). Applied aspects of seagrasses. In *Handbook of Seagrass Biology: An Ecosystem Perspective*, ed. R.C. Phillips & C.P. McRoy, pp. 297–343. New York: Garland STPM Press.

Menzies, R.J., Zaneveld, J.S. & Pratt, R.M. (1967). Transported turtle grass as a source of organic enrichment of abyssal sediments off North Carolina. *Deep-Sea Research*, **14**, 111–12.

Mitchell, C.A., Custer, T.W. & Zwank, P.J. (1994). Herbivory on shoalgrass by wintering redheads in Texas. *Journal of Wildlife Management*, **58**, 131–41.

Moncreiff, C.A., Sullivan, M.J. & Daehnick, A.E. (1992). Primary production dynamics in seagrass beds of Mississippi Sound: the contributions of seagrass, epiphytic algae, sand microflora, and phytoplankton. *Marine Ecology Progress Series*, **87**, 161–71.

Montgomery, J.L.M. & Targett, T.E. (1992). The nutritional role of seagrass in the diet of the omnivorous pinfish *Lagodon rhomboides* (L.). *Journal of Experimental Marine Biology and Ecology*, **158**, 37–57.

Morgan, M.D. & Kitting, C.L. (1984). Productivity and utilization of the seagrass *Halodule wrightii* and its attached epiphytes. *Limnology and Oceanography*, **29**, 1066–76.

Moriarty, D.J.W. & O'Donohue, M.J. (1993). Nitrogen fixation in seagrass communities during summer in the Gulf of Carpentaria, Australia. *Australian Journal of Marine and Freshwater Research*, **44**, 117–25.

Moriarty, D.J.W. & Pollard, P.C. (1982). Diel variation of bacterial productivity in seagrass (*Zostera capricorni*) beds measured by rate of thymidine incorporation into DNA. *Marine Biology*, **72**, 165–73.

Moriarty, D.J.W., Iverson, R.L., Pollard, P.C. (1986). Exudation of organic carbon by the seagrass *Halodule wrightii* Aschers. and its effect on bacterial growth in the sediment. *Journal of Experimental Marine Biology and Ecology*, **96**, 115–26.

Moriarty, D.J.W., Roberts, D.G. & Pollard, P.C. (1990). Primary and bacterial productivity of tropical seagrass communities in the Gulf of Carpentaria, Australia. *Marine Ecology Progress Series*, **61**, 145–57.

Murray, L. & Wetzel, R.L. (1987). Oxygen production and consumption associated with the major autotrophic components in two temperate seagrass communities. *Marine Ecology Progress Series*, **38**, 231–9.

Nedwell, D.B. (1984). The input and mineralization of organic carbon in anaerobic aquatic sediments. *Advances in Microbial Ecology*, **7**, 93–131.

Nelson, T.A. & Waaland, J.R. (1997). Seasonality of eelgrass, epiphyte, and grazer biomass and productivity in subtidal eelgrass meadows subjected to moderate tidal amplitude. *Aquatic Botany*, **56**, 51–74.

Newell, S.Y. (1981). Fungi and bacteria in or on leaves of eelgrass (*Zostera marina* L.) from Chesapeake Bay. *Applied and Environmental Microbiology*, **41**, 1219–24.

O'Donohue, M.J., Moriarty, D.J.W. & MacRae, I.C. (1991). Nitrogen fixation in sediments and the rhizosphere of the seagrass *Zostera capricorni*. *Microbial Ecology*, **22**, 53–64.

Opsahl, S. & Benner, R. (1993). Decomposition of senescent blades of the seagrass *Halodule wrightii* in a subtropical lagoon. *Marine Ecology Progress Series*, **94**, 191–205.

Page, H.M. (1995). Variation in the natural abundance of ^{15}N in the halophyte, *Salicornia virginica*, associated with groundwater subsidies of nitrogen in a southern California salt-marsh. *Oecologia*, **104**, 181–8.

Pedersen, M.F., Duarte, C.M. & Cebrián, J. (1997). Rates of changes in organic matter and nutrient stocks during seagrass *Cymodocea nodosa* colonization and stand development. *Marine Ecology Progress Series*, **159**, 29–36.

Peduzzi, P. & Herndl, G.J. (1991). Decomposition and significance of seagrass leaf litter (*Cymodocea nodosa*) for the microbial food web in coastal waters (Gulf of Trieste, Northern Adriatic Sea). *Marine Ecology Progress Series*, **71**, 163–74.

Penhale, P.A. (1977). Macrophyte-epiphyte biomass and productivity in an eelgrass (*Zostera marina* L.) community. *Journal of Experimental Marine Biology and Ecology*, **26**, 211–24.

Penhale, P.A. & Smith, W.O. (1977). Excretion of dissolved organic carbon by eelgrass (*Zostera marina*) and its epiphytes. *Limnology and Oceanography*, **22**, 400–7.

Penhale, P.A. & Thayer, G.W. (1980). Uptake and transfer of carbon and phosphorus by eelgrass (*Zostera marina* L.) and its epiphytes. *Journal of Experimental Marine Biology and Ecology*, **42**, 113–23.

Pereg, L.L., Lipkin, Y. & Sar, N. (1994). Different niches of the *Halophila stipulacea* seagrass bed harbor distinct populations of nitrogen fixing bacteria. *Marine Biology*, **119**, 327–33.

Pérez-Lloréns, J.L., de Visser, P., Nienhuis, P.H. & Niell, F.X. (1993). Light-dependent uptake, translocation and foliar release of phosphorus by the intertidal seagrass *Zostera noltii* Hornem. *Journal of Experimental Marine Biology and Ecology*, **166**, 165–74.

Pinto, L. & Punchihewa, N.N. (1996). Utilisation of mangroves and seagrasses by fishes in the Negombo Estuary, Sri Lanka. *Marine Biology*, **126**, 333–45.

Pollard, P.C. & Kogure, K. (1993a). The role of epiphytic and epibenthic algal productivity in a tropical seagrass, *Syringodium isoetifolium* (Aschers.) Dandy, community. *Australian Journal of Marine and Freshwater Research*, **44**, 141–54.

Pollard, P.C. & Kogure, K. (1993b). Bacterial decomposition of detritus in a tropical seagrass (*Syringodium isoetifolium*) ecosystem, measured with [methyl-^{3}H]thymidine. *Australian Journal of Marine and Freshwater Research*, **44**, 155–72.

Pollard, P.C. & Moriarty, D.J.W. (1991). Organic carbon decomposition, primary and bacterial productivity, and sulphate reduction, in tropical seagrass beds of the Gulf of Carpentaria, Australia. *Marine Ecology Progress Series*, **69**, 149–59.

Riisgård, H.U., Bondo Christensen, P., Olesen, N.J., Petersen, J.K., Møller, M.M. & Andersen, P. (1995). Biological structure in a shallow cove (Kertinge Nor, Denmark) – Control by benthic nutrient fluxes and suspension-feeding ascidians and jellyfish. *Ophelia*, **41**, 329–44.

Risgaard-Petersen, N. & Jensen, K. (1997). Nitrification and denitrification in the rhizosphere of the aquatic macrophyte *Lobelia dortmanna* L. *Limnology and Oceanography*, **42**, 529–37.

Risgaard-Petersen, N., Dalsgaard, T., Rysgaard, S., Christensen, P.B., Borum, J., McGlathery, K. & Nielsen, L.P. (1998). Nitrogen balance of a temperate eelgrass *Zostera marina* bed. *Marine Ecology Progress Series*, **174**, 281–91.

Romero, J., Pergent, G., Pergent-Martini, C., Mateo, M.A. & Regnier, C. (1992). The detritic compartment in a *Posidonia oceanica* meadow: litter features, decomposition rates, and mineral stocks. *P.S.Z.N. I: Marine Ecology*, **13**, 69–83.

Rysgaard, S., Risgaard-Petersen, N. & Sloth, N.P. (1996). Nitrification, denitrification, and nitrate ammonification in sediments of two coastal lagoons in Southern France. *Hydrobiologia*, **329**, 133–41.

Sand-Jensen, K. (1977). Effect of epiphytes on eelgrass photosynthesis. *Aquatic Botany*, **3**, 55–63.

Seitzinger, S.P. (1988). Denitrification in freshwater and coastal marine ecosystems: ecological and geochemical significance. *Limnology and Oceanography*, **33**, 702–24.

Shieh, W.Y. & Yang, J.T. (1997). Denitrification in the rhizosphere of the two seagrasses *Thalassia hemprichii* (Ehrenb.) Aschers and *Halodule uninervis* (Forsk.) Aschers. *Journal of Experimental Marine Biology and Ecology*, **218**, 229–41.

Shieh, W.Y., Simidu, U. & Maruyama, Y. (1989). Enumeration and characterization of nitrogen-fixing bacteria in an eelgrass (*Zostera marina*) bed. *Microbial Ecology*, **18**, 249–59.

Slim, F.J., Hemminga, M.A., Cocheret de la Morinière, E. & Van der Velde, G. (1996). Tidal exchange of macrolitter between a mangrove forest and adjacent seagrass beds (Gazi Bay, Kenya). *Netherlands Journal of Aquatic Ecology*, **30**, 119–28.

Stapel, J. & Hemminga, M.A. (1997). Nutrient resorption from seagrass leaves. *Marine Biology*, **128**, 197–206.

Suchanek, T.H., Williams, S.L., Ogden, D.K., Hubbard, D.K. & Gill, I.P. (1985). Utilization of shallow-water seagrass detritus by Caribbean deep-sea macrofauna: δ^{13}C evidence. *Deep-Sea Research*, **32**, 201–14.

Terrados, J. & Duarte, C.M. (1999). Experimental evidence of reduced particle resuspension within a seagrass (*Posidonia oceanica* L.) meadow. *Journal of Experimental Marine Bioogy and Ecology*, **243**, 45–53.

Thayer, G.W., Bjorndal, K.A., Ogden, J.C., Williams, S.L. & Zieman, J.C. (1984). Role of larger herbivores in seagrass communities. *Estuaries*, **7**, 351–76.

Trautman, D.A. & Borowitzka, M.A. (1999). Distribution of the epiphytic organisms on *Posidonia australis* and *P. sinuosa*, two seagrasses with differing leaf morphology. *Marine Ecology Progress Series*, **179**, 215–29.

Tresher, R.E., Nichols, P.D., Gunn, J.S., Bruce, B.D. & Furlani, D.M. (1992). Seagrass detritus as a basis of a coastal planktonic food chain. *Limnology and Oceanography*, **37**, 1754–8.

Tubbs, C.R. & Tubbs, J.M. (1983). The distribution of *Zostera* and its exploitation by wildfowl in the Solent, Southern England. *Aquatic Botany*, **15**, 223–39.

Vähätalo, A., Søndergaard, M., Schlüter, L. & Markager, S. (1998). Impact of solar radiation on the decomposition of detrital leaves of eelgrass *Zostera marina*. *Marine Ecology Progress Series*, **170**, 107–17.

Valentine, J.F. & Heck, K.L. (1991). The role of sea urchin grazing in regulating subtropical seagrass meadows: evidence from field manipulations in the northern Gulf of Mexico. *Journal of Experimental Marine Biology and Ecology*, **154**, 215–30.

Valeur, J.R. (1994). Resuspension mechanisms and measuring methods. In *Sediment Trap Studies in the Nordic Countries*, ed. S. Floderus, A. Heiskanen, M. Olesen & P. Wassmann, pp. 185–203. Helsingor: Marine Biological Laboratory.

Valiela, I., Foreman, K., LaMontagne, M., Hersh, D., Costa, J., D'Avanzo, C., Babione, M., Peckol, P., DeMeo-Andreson, B., Sham, C.-H., Brawley, J. & Lajtha, K. (1992). Couplings of watersheds and coastal waters: sources and consequences of nutrient enrichment in Waquoit Bay, Massachusetts. *Estuaries*, **15**, 443–57.

Velimirov, B. (1986). DOC dynamics in a Mediterranean seagrass system. *Marine Ecology Progress Series*, **28**, 21–41.

Velimirov, B. & Walenta-Simon, S. (1993). Bacterial growth rates and productivity within a seagrass system: seasonal variations in a *Posidonia oceanica* bed. *Marine Ecology Progress Series*, **96**, 101–7.

Verduin, J.J. & Backhaus, J.O. (2000). Dynamics of plant-flow interactions for the seagrass *Amphibolis antarctica*: field observations and model simulations. *Estuarine and Coastal Shelf Science*, **50**, 185–204.

Verhagen, F.J.M., Laanbroek, H.J. & Woldendorp, J.W. (1995). Competition for ammonia between plant roots and nitrifying and heterotrophic bacteria and the effects of protozoan grazing. *Plant and Soil*, **170**, 241–50.

Welsh, D.T., Bourguès, S., de Wit, R., Herbert, R.A. (1996a). Seasonal variations in nitrogen-fixation (acetylene reduction) and sulphate-reduction rates in the rhizosphere of *Zostera noltii*: nitrogen fixation by sulphate-reducing bacteria. *Marine Biology*, **125**, 619–28.

Welsh, D.T., Bourguès, S., de Wit, R. & Herbert, R.A. (1996b). Seasonal variation in rates of heterotrophic nitrogen fixation (acetylene reduction) in *Zostera noltii* meadows and uncolonized sediments of the Bassin d'Arcachon, south-west France. *Hydrobiologia*, **329**, 161–74.

Welsh, D.T., Bourguès, S., de Wit, R. & Auby, I. (1997). Effect of plant photosynthesis, carbon sources and ammonium availability on nitrogen fixation rates in the rhizosphere of *Zostera noltii*. *Aquatic Microbial Ecology*, **12**, 285–90.

Wijsman, J.W.M., Herman, P.M.J., Middelburg, J.J. & Soetaert, K. (2000). A model for early diagenetic processes in sediments of the continental shelf of the Black Sea. *Estuarine, Coastal and Shelf Science* (in press)

Winding, A. (1992). [^{3}H]thymidine incorporation to estimate growth rates of anaerobic bacterial strains. *Applied and Environmental Microbiology*, **58**, 2660–2.

Worcester, S.E. (1995). Effects of eelgrass beds on advection and turbulent mixing in low current and low shoot density environments. *Marine Ecology Progress Series*, **126**, 223–32.

Wolff, T. (1979). Macrofaunal utilization of plant remains in the deep-sea. *Sarsia*, **64**, 117–36.

Ziegler, S. & Benner, R. (1998). Ecosystem metabolism in a subtropical, seagrass-dominated lagoon. *Marine Ecology Progress Series*, **173**, 1–12.

Ziegler, S. & Benner, R. (1999). Dissolved organic carbon cycling in a subtropical seagrass-dominated lagoon. *Marine Ecology Progress Series*, **180**, 149–60.

6

Fauna associated with seagrass systems

6.1 Introduction

The meadows formed by seagrasses have characteristics that make them a suitable habitat for many species of animals. The high primary productivity of the seagrasses, augmented with that of epiphytic and benthic algae, ensures an abundant supply of organic matter that can be used as the basic energy source for more or less complicated food webs. Moreover, the three-dimensional structure of the vegetation, with its network of roots and rhizomes and often dense leaf canopy, offers hiding places that protect against predation, and also provides substrate for attachment. The vegetation structure, furthermore, confers physical and chemical qualities to the environment that may attract fauna: currents within the canopy are reduced, the sediment is stabilized and often fine grained, and irradiance conditions are modified. In this chapter we will first take a closer look at the general abundance and species richness of the fauna associated with seagrass meadows, before turning to the faunal groups that have received major attention, i.e. fishes, crustaceans and molluscs. The association of sea cows and turtles with seagrass beds will also be discussed. The significance of seagrass meadows as a habitat and foraging area is a recurrent theme in these sections. In the final part of the chapter, the ways in which the fauna affect the functioning of the seagrasses will be addressed.

6.2 Abundance and diversity

The fauna of seagrass meadows are heterogeneous assemblages of animals belonging to a variety of taxa, with many different ecological characteristics. With respect to the compartment of the system where the animal is living, several broad categories can be distinguished: (1)

infaunal species, animals that live in the sediment; (2) epifaunal species, the animals living on the stems and leaves, which includes both sessile and motile animals associated with the plant surfaces, and the animals living on the sediment surface; and (3) epibenthic species, larger mobile animals that are loosely associated with the seagrass beds, moving freely under and over the leaf canopy (e.g. fishes).

The question of whether seagrass meadows support a greater diversity and abundance of animals than adjacent areas without seagrass vegetation has received ample attention over the years. Usually, these studies focus on only one or two of the above-mentioned faunal categories, one of the reasons being that collecting the animals belonging to the different categories requires quite different sampling methodologies. For the mobile species, for instance, nets are commonly used, whereas sampling the infauna requires soil corers. The studies also vary considerably with respect to the intensity of sampling efforts: from a single sampling effort to a programme of repeated sampling over periods of one or several years. Despite the variation in scope of the studies and in sampling strategies, a fairly clear picture emerges. This is illustrated by the compilation of data from 24 recent studies on abundance and species richness in seagrass systems and other systems (Table 6.1). In most of these studies, seagrass meadows have been compared with bare areas. In addition, several comparisons with other systems have been made, notably with mangrove areas, systems that often occur adjacent to seagrasses in tropical regions. We can draw the following conclusions from the data: (1) seagrass meadows harbour, virtually without exception, more animals and more species than nearby unvegetated areas; this is generally true for all the broad faunal groups that were studied, whether it concerned fish, decapods, or benthic fauna; (2) in comparison with other habitats, the results are less straightforward. The first conclusion by and large holds when the seagrass meadows are compared with unvegetated zones in other systems (e.g. saltmarsh creeks, mudflat channels) and to open water, but if other vegetated areas are considered (*Ulva* mats, reef-algal areas, saltmarshes, mangroves), animal numbers and species diversity are repeatedly higher in these areas than in seagrass meadows. It thus seems that the fauna associated with seagrass beds is abundant and diverse when contrasted with bare sediments, but that animal abundance and species diversity is not particularly high in comparison with other macrophyte habitats in the marine coastal zone.

The faunal community structure of seagrass meadows is not a fixed or constant attribute for any particular seagrass species or combination of

Table 6.1. *Abundance and species richness of animals in seagrass systems*
Unless indicated otherwise, the numbers pertain to cumulative numbers of individuals and species collected over the duration of the respective studies.

Species	Numbers in seagrass system	Numbers in nearby unvegetated area	Numbers in nearby other system
Fish			
Z. marina[1]	individ.: 4560 species: 14	individ.: 147 species: 7	
Z. marina[2]	individ.: 15.7 (av./sample) species: 12	individ.: 8.2 (av./sample) species: 12	*Marsh creeks* individ.: 51.7 (av./sample) species: 11
Z. capricorni[3]	individ.: 41 141 species: 49	individ.: 2251 species: 29	
Z. capricorni[4]	individ. (day): 7824 species (day): 32 individ. (night): 9396 species (night): 39	individ. (day): 380 species (day): 16 individ. (night): 3884 species (night): 20	
Z. muelleri / H. tasmanica[5]	individ.: 9866	individ.: 4005	
H. tasmanica[6]	individ.: 189 (av./sample) species: 33	individ.: 50.1 (av./sample) species: 28	*Ulva mats* individ.: 4.35 (av./sample) species: 12
Z. muelleri[6]	individ.: 60.5 (av./sample) species: 24	individ.: 100.9 (av./sample) species: 16	
H. tasmanica[7]	individ.: 45.1 (av./sample) species: 28	individ.: 5.5 (av./sample) species: 20	*Reef-algal habitat* individ.: 19.4 (av./sample) species: 31

Table 6.1 (*cont.*)

Species	Numbers in seagrass system	Numbers in nearby unvegetated area	Numbers in nearby other system
H. tasmanica[7]	individ.: 51.8 (av./sample) species: 30	individ.: 20.5 (av./sample) species: 17	*Reef-algal habitat* individ.: 35.4 (av./sample) species: 27
Z. muelleri / H. tasmanica[8]	individ.: 108.6 (av./sample) species: 28	individ.: 11.9 (av./sample) species: 21	*Mudflat channels* individ.: 9.0 (av./sample) species: 26 } collected with seine
Z. muelleri / H. tasmanica[8]	individ.: 374 species: 23	individ.: 742 species: 19	individ.: 150 species: 21 } collected with gillnets
Mixed beds[9]	individ. (July): 4.4 individ. (Sept): 4.6 (no./ m²)		*Mangrove* individ. (July): 6.9 individ. (Sept): 11.1 (no./ m²) *Open water* ind. (July): 2.54 ind. (Sept): 10 (no./m²)
Mixed beds[10]	species: 30	species: 25	*Mangrove* species: 40 *Mangrove fringe* species: 36
H. ovalis / H.pinifolia[11]	species: 46		*Mangrove* species: 56
T. testudinum[12]	individ.: 477 species: 58	individ.: 374 species: 46	*Mangrove creek with T. testudinum* individ.: 2445 species: 74
T. testudinum[13]	individ.: 1408 species: 17	individ.: 185 species: 26	
H. wrightii / R. maritima[15]	individ. (Sept.): 10.1 individ. (May): 9.0 (no./m²)	individ. (Sept.): 1.9 individ. (May): 2.1 (no./m²)	*Salt marsh* individ. (Sept.): 8.7 individ. (May): 11.2 (no./m²)

	Seagrass (1)	Seagrass (2)		
Crustaceans				
H. wrightii / R. maritima[15]	individ. (Sept.): 23.4 individ. (May.): 47.7 (no./m²)	individ. (Sept.): 1.8 individ. (May.): 1.3 (no./m²)	*Salt marsh* individ. (Sept.): 130.5 individ. (May.): 270.8 (no./m²)	*Marsh creeks* individ.: 59.0 (av./sample) species: 5 *Open water* ind. (July): 32.5 ind. (Sept): 49 (no./m²)
Decapods				
Z. marina[14]	individ.: 2117 species: 8	individ.: 355 species: 6		
Z. marina[2]	individ.: 73.3 (av./sample) species: 6	individ.: 22.7 (av./sample) species: 6	*Ulva mats* individ.: 55.2 (av./sample) species: 10	
Mixed beds[9]	individ. (July): 100.1 individ. (Sept): 149.0 (no./m²)		*Mangrove* individ. (July): 6.3 individ. (Sept): 7.8 (no./m²)	
Penaeid prawns				
Z. capricorni[16]	individ.: 42 (av./sample)	individ.: 8.9 (av./sample)		
Canopy epifauna				
Z. muelleri[17]	individ.: 9322	individ.: 1639		
Canopy and benthic macrofauna				
Z. marina[18]	individ.: 19 095 species: 83	individ.: 9969 species: 62	*Marsh channel* individ.: 4114 species: 30	*Marsh pool* individ.: 3201 species: 17
Benthic macrofauna				
Z. muelleri / H. tasmanica[19]	species: 300	species: 185	*Subtidal channel* species: 265	
Mixed beds[20]	individ.: 4023 (no./m²)	individ.: 815 (no./m²)		

Table 6.1 (*cont.*)

Species	Numbers in seagrass system	Numbers in nearby unvegetated area	Numbers in nearby other system
Z. marina [21]	individ.: 23 500–52 000 (no./m²) species: 5.9–8.8 (no./m²)	individ.: 2500–16 000 (no./m²) species: 2.2–5.5 (no./m²)	
Mixed beds [22]	individ.: 3865 species: 193	individ.: 3839 species: 155	*Mangrove* individ.: 9850 species: 87
Benthic polychaetes			
P. oceanica [23]	species: 161 individ.: 598	species: 69 individ.: 33	*Rocky bottom* species: 76
P. oceanica [24]	species: 50	species: 12	

(1) Heck *et al.*, 1989. (2) Sogard & Able, 1991. (3) Gray *et al.*, 1996. (4) Gray *et al.*, 1998. (5) Connolly, 1994a. (6) Jenkins *et al.*, 1997a; only data from North Jetty and Queenscliff are shown. (7) Jenkins & Wheatley, 1998; only data from St. Leonards and Grassy Point are shown. (8) Edgar & Shaw, 1995a; only data from Peck Point are shown. (9) Sheridan, 1992. (10) Laegdsgaard & Johnson, 1995. (11) Pinto & Punchihewa, 1996. (12) Sedberry & Carter, 1993. (13) Jordan *et al.*, 1996. (14) Heck *et al.*, 1989. (15) Rozas & Minello, 1998; densities do not include uncommon species with < 10 individuals. (16) Coles *et al.*, 1993; data pertain to commercial penaeid prawn species only. (17) Connolly, 1997. (18) Heck *et al.*, 1995. (19) Edgar *et al.*, 1994. (20) Ansari *et al.*, 1991. (21) Boström & Bonsdorff, 1997. (22) Sheridan, 1997. (23) Somaschini *et al.*, 1994. (24) Gambi *et al.*, 1998.

species. A recent illustration of this can be found in a study of the benthic macrofauna in *Zostera marina* beds in the Baltic Sea (Boström & Bonsdorff, 1997). In five different beds (10–270 km apart) a total of 28 infaunal species were recorded. Similarity between sites, expressed by the Jaccard coefficient (= *a*/*b*, where *a* is the number of species common to the two sites and *b* is the total number of species found at the two sites) ranged between 0.32 and 0.72. For the *Zostera* epifauna that was studied at four localities, and that was represented by 27 species, similarity values ranged between 0.37 and 0.65. Another example from another part of the globe is given by a study of the macroinvertebrate fauna occurring in three mixed meadows of *Heterozostera tasmanica* and *Zostera muelleri* along the coast of central Victoria (Australia). In these meadows, where a diverse benthic fauna comprising hundreds of species was collected, the faunal similarity between the meadows, again expressed using Jaccard's coefficient, was only ca. 0.5 (Edgar *et al.*, 1994). Striking differences in the community structure of fish species residing in seagrass beds of similar species composition have also been noted (Heck *et al.*, 1989; Ferrell *et al.*, 1993; Francour, 1997).

There are several reasons for this variability in faunal assemblages. In the first place, the physico-chemical environment may vary and differ between seagrass growth locations (e.g. with respect to hydrodynamical conditions, salinity and depth). These factors may be of overriding importance for the presence or absence of a species, whether there is seagrass vegetation or not. This becomes more obvious if we consider the fact that only a few animal species are exclusively dependent on sea-grasses, such as a few species of southern Australian chitons that appear to be restricted to particular seagrass genera (Howard *et al.*, 1989), the Mediterranean fan mussel, *Pinna nobilis*, that lives in *Posidonia oceanica* meadows (Richardson *et al.*, 1999), and the specialist isopod leaf borers of the genus *Lynseia* that live in specific seagrass species (Brearley & Walker, 1995). Most species, however, also occur outside seagrass meadows, and have no strict binding to seagrasses. The presence of these species does not crucially depend on seagrass vegetation, but is deter-mined by other factors. Secondly, the variable geographical setting of seagrass meadows as part of the larger coastal system also is a factor of importance in causing variability in seagrass faunal assemblages. Among other things, this is because many of the animals found in seagrass beds have pelagic larvae that recruit to the beds; the geographical location of the beds relative to spawning locations thus determines whether the larvae can easily reach the beds with the prevailing currents or not.

Thirdly, many mobile species spend only a part of their life cycle in the seagrass bed. Distinct settlement episodes and the more or less synchronized departure of individuals at a certain life stage may result in alternating periods of presence and absence of a species.

6.3 Fish assemblages

6.3.1 The foraging and refuge function of seagrasses

The fish fauna of seagrasses can be of a considerable diversity. The occurrence of more than 100 species associated with seagrass beds in a certain region is no exception. These fishes can be classified, in a modification of the scheme developed by Kikuchi (1966), into four categories depending on their residence status: *permanent residents*, species that stay throughout their life in the seagrass habitat; *temporary residents*, species that are only seasonally or during a part of their life cycle present in the meadows; *regular visitors*, species that frequently visit the meadows, e.g. fishes that migrate on a diurnal basis from coral reefs to nearby seagrass beds, and *occasional visitors*, fishes that visit the beds only now and then. As can be gathered from Table 6.1, there are generally more fish in seagrass beds than in bare areas. Why is this the case? Enhanced food resources relative to unvegetated substrates and protection from predators have been generally considered to be the main benefits of seagrass beds for fish. Direct herbivory on seagrass is relatively uncommon among the members of the seagrass fish communities, as we have seen earlier (section 5. 3). The standing crop of seagrass plants is thus a rather unexploited source of organic matter as far as the fishes are concerned, and other types of food are more important. The diet of fish in seagrass habitats has been the subject of exhaustive earlier reviews (Pollard, 1984; Klumpp *et al.*, 1989). The main conclusions reached in these reviews have not been fundamentally changed by more recent studies. In general, small invertebrates are of overriding importance as prey items for the fish fauna. For most of the fish, planktonic and epifaunal crustaceans are the dominant components of their diet. Among the crustaceans, amphipods, copepods and shrimps have emerged as the most important food categories. Infaunal benthos usually is of lesser importance, although the consumption of bivalve molluscs, polychaetes and other soil dwelling fauna certainly is not rare (Edgar & Shaw, 1995c; Motta *et al.*, 1995). Small fish residing in the seagrass canopy may also be a prey itself. Diet shifts with growth commonly occur in fish, and at more advanced growth stages fish may start to prey on other, smaller fishes, thereby adding a new component to their diets. True piscivores, however, generally appear to be few in numbers among the seagrass-associated fishes.

Small crustaceans, the favoured food of seagrass fish fauna, can be abundantly present in seagrass meadows. In a study of the trophic ecology of the fish fauna of vegetated and unvegetated habitats at 14 sites along the southern Australian coast, the estimated production of crustaceans was much higher in the seagrass habitat than in unvegetated sites in nearly all cases (Edgar & Shaw, 1995c), which may have been the reason for the higher fish densities that were present at the seagrass sites. The importance of crustaceans as a food source was evident in this study from straightforward observations on gut contents, and also by the finding that the fishes virtually completely consumed the crustacean production, whereas only a small proportion of the non-crustacean benthic production was consumed. Furthermore, fish production was highly correlated with crustacean production.

It is well known that in situations where seagrasses adjoin reefs, fish show diurnal migrations from the reefs to forage over the seagrass beds (Weinstein & Heck, 1979; Robblee & Zieman, 1984; Baelde, 1990). Although such observations demonstrate that the supply of food can be important in attracting fish to seagrass meadows, food certainly is not the only relevant factor explaining the presence or abundance of fish in the seagrass habitat. For many fish, the refuge function of the seagrass canopy is likely to be the main reason. This can be illustrated by a growth study of three fish species in New Jersey estuaries: *Pseudopleuronectes americanus* (winter flounder), *Tautoga onitis* (tautog), and *Gobiosoma bosci* (naked gobi). Juveniles of these species are associated with vegetation (eelgrass, *Zostera marina*, or the macroalga *Ulva lactuca*) and also (except *T. onitis*) occur on bare sand. Growth of these species was followed while they were kept in cages on sites with and without seagrass or macroalgae. It appeared that *P. americanus* and *T. onitis* grew faster at sites where they are normally most abundant, either at sites with or without plants. *G. bosci*, however, showed better growth on bare sand than on eelgrass sites, although the juveniles of this species normally achieve their highest densities in the latter habitat (Sogard, 1992). Apparently, the food supply in eelgrass is suboptimal for this species, and it sacrifices part of its growth potential in order to gain another benefit associated with the vegetation. A plausible explanation for this finding is that reduced rates of predation underlie the selection of the eelgrass habitat.

A number of studies have shown that the predation risk to invertebrates and fish is diminished when they reside in the canopy. For a key review of the factors that determine predator–prey interactions in seagrass beds, the interested reader is referred to Orth *et al.* (1984). The complexity of the habitat, expressed by vegetation parameters such as shoot density or leaf surface area, is important in this respect: with increasing habitat complexity the foraging success of predators is reduced. This relationship is non-linear, and many studies have indicated that a threshold level of complexity is required before the foraging success of the predator is significantly reduced (Gotceitas & Colgan, 1989). Behavioural studies on juvenile fish and other prey animals, furthermore, have demonstrated that, confronted with predators, the prey animals respond by leaving bare areas and seek refuge in more complex habitats. Gotceitas & Colgan (1989) pointed out that if some threshold level in habitat complexity exists before predation success is effectively reduced, it can also be predicted that prey animals select habitats above a certain threshold level of complexity to avoid predation. Their experiments with artificial vegetation and lake fish (juvenile bluegill sunfish and predatory largemouth bass) indeed clearly show both the existence of a non-linear relationship between plant stem density and foraging success (with an estimated threshold level of 276 stems per m^2), and a non-linear relationship between stem density and refuge choice by the prey (with an estimated threshold level of 516 stems per m^2; Fig. 6.1).

The protection against visually oriented predators provided by seagrass will be primarily due to restricted prey visibility; furthermore, the manoeuvrability of any predator will be influenced by the shoots and leaves projecting up into the water column, affecting its foraging success. For different combinations of prey, predator, and vegetation type, variability in the relationship between habitat complexity and predation success thus can be expected, depending on factors such as plant morphology, prey coloration and size, and predator size and agility.

6.3.2 Seagrasses as nursery habitat

Juvenile specimens often dominate the fish communities of seagrass beds. In a study of the fish communities of estuarine seagrass beds in northern Queensland (Australia), for instance, 134 species were recorded. The average length of the individual specimens caught was only 32 mm, because most of the fishes were juveniles, although some large adult

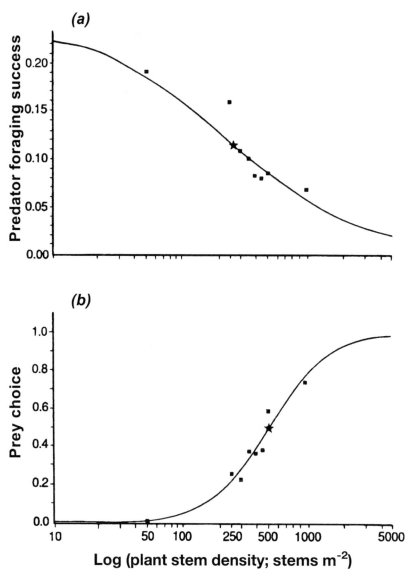

Fig. 6.1. Relationship between stem density of artificial vegetation and (*a*) foraging success of largemouth bass (*Micropterus salmoides*) preying on juvenile bluegill sunfish (*Lepomis macrochirus*), and (*b*) prey habitat choice. The star symbol indicates the estimate of the 'threshold level' of plant stem density necessary to affect the relationship significantly; black squares indicate observed values. (Gotceitas & Colgan, 1989.)

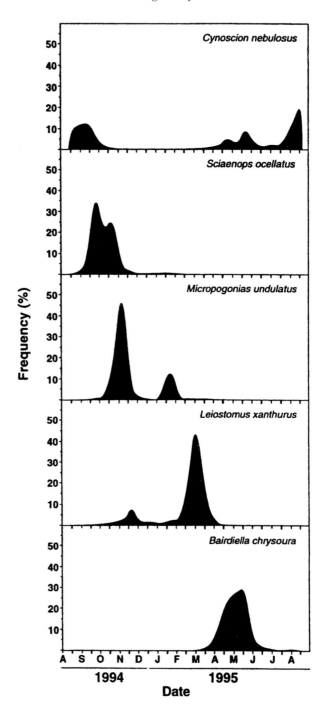

fishes with lengths over 60 cm were also present (Coles *et al.*, 1993). Such observations have led to the widely held notion that seagrass meadows are nursery habitats for many fishes. Indeed, all over the world seagrass meadows are exploited by fish species in their early life stages. Spawning of these species usually does not occur within the beds, but takes place elsewhere, e.g. in the coastal ocean. Settlement into the seagrass meadows follows after a pelagic larval phase. The juveniles remain in the beds only for a certain period, which commonly lasts several weeks or months. Upon reaching a critical length, they leave the beds and move to other habitats. The recruitment of juvenile fish to seagrass beds may occur in discrete pulses. Because the growth stage upon which they leave the beds is also reached in more or less the same time, many species are only abundantly present at intervals. Examples of this pattern are shown in Fig. 6.2 (Rooker *et al.*, 1998). Among the number of fish species that may spend part of their juvenile existence in seagrass beds, there are several species that are important in recreational and commercial fisheries. Examples can be found among the Serranidae (epinepheline groupers), whose adult populations are found near offshore reefs and rocky bottoms (Ross & Moser, 1995), among the Sciaenidae (drums), and among the Sillaginidae (whiting), which live as adults in deeper inshore and offshore waters (Hyndes *et al.*, 1996; Rooker *et al.*, 1998). Large stocks of adult Atlantic cod (*Gadus morhua*) can be found in northern parts of the Atlantic Ocean; some of them may have once used eelgrass beds as a nursery area when they were juveniles (Gotceitas *et al.*, 1997).

Although the fact that seagrass beds are nursery habitats for many species is firmly established, the exclusivity of this feature has somewhat faded over the years with the growing number of studies that show that juveniles known to occur in seagrass beds may also use alternative nursery habitats. Juveniles of the same species thus may utilize both seagrasses and reef-algal habitats as nursery areas (Jenkins & Wheatley, 1998), or seagrass beds and oyster rocks (Ross & Moser, 1995), or seagrass and sea lettuce (*Ulva lactuca*) mats (Sogard & Able, 1991), or they may even use bare sediments as an alternative to seagrass beds (Blaber *et al.*, 1992; Edgar & Shaw, 1995a). Such alternative habitats present in the same region may be of superior value as a nursery habitat.

Fig. 6.2. Relative abundance of sciaenid fishes (with lengths ≤ 40 mm) collected from seagrass meadows in the Aransas Estuary, USA. Percentage frequency estimates based on the number of individuals per sampling date. (Rooker *et al.*, 1998.)

Laegdsgaard & Johnson (1995), for instance, showed that the abundance of many juvenile fish species was much higher in mangrove forests than in adjacent seagrass meadows. Compared with the mangrove forests, the nursery function of the seagrasses in this area (south-east Queensland, Australia) for fish thus seemed to be limited.

Finally, it is important to note that seagrass meadows may vary considerably in terms of nursery value, even if the same species is considered. Eelgrass (*Zostera marina*) meadows in the more northern part of the west Atlantic seem to be less important as nursery area for commercially important fish species than eelgrass meadows of the mid-Atlantic coast (Heck *et al.*, 1989). In Mediterranean meadows of *Posidonia oceanica*, only the more shallow beds (<15 m), not the deeper ones, appear to be important as a nursery area for fishes (Francour, 1997). Such differences may be due to several factors that influence the species composition and abundance of juveniles, or that are relevant to the fish community as a whole. These factors are discussed in the next section.

6.3.3 Determinants of faunal composition and abundance

The wealth of studies on fish assemblages associated with seagrass meadows show that great variability exists in fish species diversity and abundance among seagrass meadows. At least four factors are relevant to this variability: (1) the vegetation structure of the beds, (2) the extent of larval and juvenile settlement in the meadows, and postsettlement mortality and migration processes, (3) the location of the beds relative to other fish habitats, and, (4) the physico-chemical environment of the beds.

Among the seagrass species, a variety of leaf shapes and leaf lengths occurs. This variety, intensified in the field situation by local differences in shoot density and by the variable presence of co-occurring macroalgae, brings on a differentiation in habitat value among seagrass systems. Various whiting species, for instance, shun the *Posidinia australis* and *P. sinuosa* beds that dominate the seagrass vegetation in the nearshore waters of the lower west coast of Australia (Hyndes *et al.*, 1996). Juvenile whiting, however, do use meadows formed by *Zostera muelleri*, *Z. capricorni* and *Heterozostera tasmanica* in marine embayments in eastern Australia (Hyndes *et al.*, 1996; Connolly, 1994a, Burchmore *et al.*, 1988). This discrepancy may arise from the far denser canopy of *Posidonia* species that hampers penetration and movement of the whiting (Hyndes *et al.*, 1996).

Studies *within* individual seagrass beds generally show that more animals occur in areas with higher shoot densities. The relatively low numbers of animals in sparsely vegetated areas may be partly the direct consequence of a higher predation level, but this is not necessarily so. The low numbers may also be simply the consequence of habitat preference, leading to the selective avoidance of less dense areas, as field experiments suggest (Bell & Westoby, 1986a). Predation, of course, may be the ultimate selective force leading to such behaviour, but the reduced availability of food may be an alternative reason (Connolly, 1994b).

The general increase in total faunal abundance with increasing shoot density may not consistently hold for every individual species. A study of the effects of experimental manipulation of *Zostera capricorni* canopies on fish and decapods showed that thinning led to a decrease in abundance of seven species, but also to an increase in three species, whereas the abundance of other species was not affected at all. Similar mixed results were obtained after manipulations of *Posidonia australis* vegetation (Bell & Westoby, 1986b). Relationships between faunal numbers and shoot density may also not be evident when the pooled data from separate meadows occurring over a wider spatial scale are considered. The precariousness of recruitment to the seagrass habitat may be the reason for this.

Recruitment of larvae and early juveniles to the seagrass habitat is a factor that directly bears on the presence and abundance of the fish fauna. As mentioned, few species actually reproduce among seagrasses; instead, spawning usually occurs away from the seagrass beds. The function of seagrass beds as a site of reproduction is probably largely confined to the permanent residents. Fish belonging to this category are often represented by small, unconspicuous species that live under the canopy. These small resident species are likely to have generally short life spans, which necessitates continually successful recruitment. Sogard *et al.* (1987) made the interesting observation that the resident fishes of the seagrass meadows in their study had reproductive strategies that minimized planktonic dispersal, from attached eggs and parental brooding to complete elimination of a pelagic stage. The reason for this phenomenon may be that such strategies increase the chances of continuous recruitment of juveniles to seagrass beds where the adults were able to spawn successfully.

The preponderance of non-local spawning makes the distribution of the seagrass-associated fish fauna critically dependent on the supply and settlement of planktonic larvae (or, in some instances, of planktonic early juveniles). Variable settlement is now commonly acknowledged as an important factor in explaining the different patterns in diversity and abundance of the fish fauna of seagrass meadows, particularly due to the hypothetical model of Bell & Westoby (1986c), which proved to be a considerable impetus for further studies. Formulation of this model was inspired by the observation that, although the physical complexity of a seagrass canopy usually affects the abundance of associated fish and decapods *within* a bed, on a wider scale (*among* beds in the same region) correlations between complexity and abundance are rare. Bell & Westoby tried to reconcile these findings by assuming the following three processes: (1) larvae of fishes and decapods settle into the first seagrass bed they encounter, irrespective of its physical complexity; spatial and temporal variations in larval supply are thus reflected in a variable distribution of postsettlement larvae and juveniles among beds; (2) after settling, individuals redistribute within the bed, selecting the type of cover that favours survival; (3) sedentary species remain in the beds, whereas mobile species, once they are large enough, may move to other beds.

In the literature the model is often briefly referred to as the 'settle and stay' model, although this term disregards the potential migration of mobile species that is acknowledged in the model. Subsequent studies have shown that many observations on species presence and abundance are in agreement with the model, although the complexity of nature, as always, is a guarantee for differences between species and situations. The question of whether or not settlement of larvae is influenced by structural characteristics of seagrass meadows is still a matter of controversy. The contribution of the variability in larval supply to spatial and temporal differences in recruitment, however, is beyond dispute. This can be conveniently illustrated by the recent study of Hannan & Williams (1998). These authors investigated the juvenile fish species occurring over *Zostera capricorni* beds in an Australian coastal barrier lagoon with one entrance channel to the Pacific Ocean. The total number of species fell with distance from the entrance, from 61 species near the entrance to 36 at distances of more than 30 km from the entrance. This decline was largely due to the decrease in number of ocean-spawning species, from 30 near the entrance to a mere 10 at the distant sites (Fig. 6.3). Apparently, the distance from the ocean in combination with restricted water circulation in the lagoon limits larval dispersal and, hence, leads to limited juvenile recruitment to seagrass beds further away from the

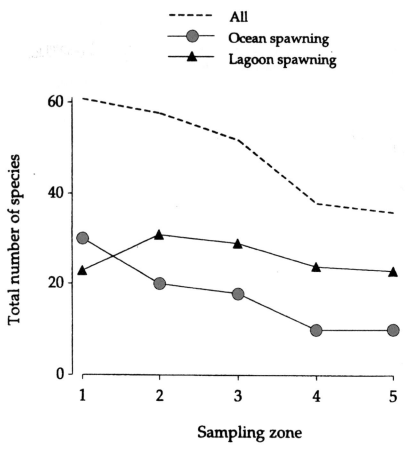

Fig. 6.3. Juvenile fish associated with seagrass meadows in a coastal barrier lagoon (Lake Macquarie) in south-east Australia.Total number of species (dashed line), ocean spawning species (shaded circles), and lagoon-spawning species (black triangles). Sampling zones 1–5 indicate increasing distance to the lagoon's entrance. (Hannan & Williams, 1998.)

oceanic source of the larvae. Likewise, the variation in availability of pelagic larvae that are ready to settle presumably is an important factor in the variable abundance of juvenile estuarine fish at different distances from the estuarine mouth.

Recent studies have demonstrated that initial settlement patterns do not necessarily persist, but may be rapidly modified due to local physical processes and behavioural patterns in the postsettlement phase.

Temporal variability in the supply of larvae thus can be clearly reflected in short-term recruitment variability, but over longer time scales a decoupling of the relationship between larval supply and abundance of recruits may occur (Hamer & Jenkins, 1996). Exposure of meadows to wave action or strong currents may lead to large postsettlement losses of larvae and juveniles (Jenkins *et al.*, 1997b), whereas settlers may also rapidly migrate to other, more preferred habitats shortly after arrival (Jenkins *et al.*, 1996; Tolan *et al.*, 1997; Rooker *et al.*, 1998). Migration movements of older individuals further advance redistribution of the fish fauna. Mobility of this type is probably common; the fact that newly available habitat (e.g. restored seagrass meadows) is rapidly colonized by fish and other fauna undoubtedly can be partly attributed to the emigration of older individuals from other habitats, as well as larval settlement from the plankton.

The proximity of other ecosystems that are themselves a habitat for fish has a strong influence on the composition and abundance of fish in the seagrass habitat. This is particularly well documented for seagrasses near coral reefs. Many fishes that are associated with the reefs during the day time migrate to the seagrass meadows at night to find food, or just to rest in the shelter of the seagrass canopy. These nocturnal invasions of reef fish enhance species diversity in the seagrass habitat. The fish assemblages collected in multispecies seagrass beds adjacent to coral reefs in Madagascar present an extreme example of this phenomenon. In these beds, the record number of 189 species was collected; many of these species were primarily coral reef associated (Pollard, 1984). Not only the proximity of coral reefs, but also that of mangroves affects the fish fauna in seagrass beds. *Thalassia testudinum* beds in Guadeloupe, which were either located close to coral reefs or close to the shore in a bay with extensive mangrove forests, harbour distinct fish assemblages. In the beds near the mangroves the species were predominantly (75%) estuarine-associated species, whereas the larger species number found in the sea-grasses near the coral reefs consisted of only 15% of estuarine-associated species. The presence of the mangroves also influenced the utilization of the beds. These beds appeared to function primarily as an additional nursery ground to the mangroves for juveniles of estuarine and pelagic species (Baelde, 1990).

Seagrass fish communities are, finally, also structured by physical characteristics of the environment. The fishes in eelgrass meadows on Cape Cod, Massachusetts, for instance, are dominated by cold-temperate species, and differ from fish assemblages in eelgrass beds from more southerly latitudes (Heck *et al.*, 1989). In tropical estuarine seagrass beds in Sri Lanka, the number of fish species significantly decreased with water temperature (Pinto & Punchihewa, 1996). On the other hand, high temperature stress may cause a low species diversity in shallow lagoons (Pollard, 1984).

Other physical factors that are of importance are salinity, exposure of the bed, and sediment structure (Sogard *et al.*, 1987; Pinto & Punchihewa, 1996). Such factors also influence the structure and diversity of seagrass vegetation, and therefore the direct effects of these physical factors on the fish fauna undoubtedly may be compounded by their effects on canopy characteristics. Changes in fish communities with depth have also been repeatedly observed (e.g. Francour, 1997). Again, in this case it is difficult to tell whether the latter changes are directly caused by a change in the abiotic environment, or are the result of biotic factors that change with depth, such as vegetation structure and larval supply.

6.4 Crustaceans

6.4.1 *General*

Like most other marine benthic systems, seagrass meadows are inhabited by numerous crustaceans. Representatives belonging to a variety of taxa and forms are present. Some of them are sediment dwellers: copepods, ostracods and amphipods are dominating meio- and macrobenthic taxa in this respect. Others live on the sediment surface or are associated with the seagrass canopy, being part of the epifauna. Among them are isopods, and again many copepods, ostracods, and amphipods. In addition, mobile decapods such as prawns, crabs and lobsters live between the shoots. These animals can form a conspicuous component of the crustacean fauna in seagrass systems. Crustaceans utilize different food sources available in the seagrass beds. Some of them, notably isopods and crabs, consume living seagrass tissue. Direct consumption of plant parts is not necessarily restricted to leaves only: a study of the diet of the omnivorous crab *Macrophthalmus hirtipes* showed that nearly half of the amount of the seagrass tissue present in its foregut consisted of root and rhizome material (Woods & Schiel, 1997). Alphaeid shrimps in tropical waters also have been reported to harvest large quantities of living seagrass. These animals clip off living leaves from the shoots and transport them into their burrows in between the plants, probably for later consumption. In an Indonesian *Thalassia hemprichii* meadow, the leaf-harvesting activity of these shrimps accounted for approximately 40% of total leaf productivity (Stapel, 1997). Predation on other animals in the seagrass beds also occurs. This may include the capture of other crustaceans, even conspecifics (Olmi & Lipcius, 1991; Moksnes *et al.*, 1997). Most crustaceans, however, particularly the smaller specimens, use algae or detrital particles as a food source (Klumpp *et al.*, 1989). The periphyton layer on the leaves – a complex of small unicellular and filamentous algae, detritus, bacteria and sediment particles – and the microflora and detrital particles on the sediment surface hence are utilized as food sources by many crustacean species. It is not unthinkable that competitive interactions might arise from limiting levels of these resources. Although microalgae and detritus may appear abundant in the seagrass habitat, their actual availability as food may be restricted by the capabilities of the various crustacean species to handle them. The specific morphology and toughness of the fragments, their dimensions relative to the size of the crustaceans, and the mouthpart arrangement of the animals all could be relevant in this respect. Manipulation of seagrass communities with the aim of investigating the role of competition in controlling epifaunal densities (including that of epifaunal crustaceans) is consistent with the idea that microalgal food limitation may occur. In confined field plots where 80% of the epifauna was reduced, the abundances of virtually all species rapidly increased, whereas in plots in which the production of epiphytes and seagrass was reduced by screening the plots from light, epifaunal number declined (Edgar, 1990).

Crustaceans are a major source of food to the fishes foraging in seagrass beds. Primarily via the food chain that links microalgae and detritus through epifaunal crustaceans to smaller fish, and the latter to larger fish predators (Edgar & Shaw, 1995b), the crustaceans are thought to play a prominent role in the energy transfer from primary producers to higher trophic levels. Crustaceans are very important in yet another aspect: several decapod species that occur in seagrasses (at least during part of their life cycle) are highly valued as human food. These decapods are the basis for extensive fisheries, particularly in warm-temperate and tropical areas. Examples are the grooved tiger prawn (*P. semisulcatus*), which is caught commercially throughout the Indo-West Pacific region, the brown tiger prawn (*P. esculentus*) which is endemic to Australia, and pink shrimp (*Penaeus duorarum*), blue crabs (*Callinectes sapidus*) and dungeness crabs (*Cancer magister*) that are fished in the western Atlantic. It is because of this commercial value that among the crustaceans associated with seagrasses the decapods have received most scientific attention.

6.4.2 Seagrasses as nursery habitat for decapods

Usually, decapod crustaceans have a pelagic larval phase, after which they settle as postlarvae and adopt a benthic life style. Seagrass beds have long been recognized as one of the coastal habitats where settlement and growth of postlarvae and juveniles may occur (e.g. Orth *et al.*, 1996). Some decapods may be found up to the adult phase, but frequently the seagrasses are used as a nursery habitat by postlarvae and juveniles only. The importance of the seagrass habitat as a nursery habitat relative to other benthic environments, however, differs widely between species. This can be illustrated by comparing different species of penaeid prawns. These prawns generally breed in offshore water. The planktonic larvae are transported inshore within 2 to 3 weeks and settle in coastal and estuarine nursery grounds. After a growth period of several weeks to months, the animals emigrate from their nursery habitats to offshore areas that may be as as much as 100 km away. In the nearshore waters of Australia, juvenile tiger prawns (*Penaeus esculentus* and *P. semisulcatus*) are predominantly found in seagrass meadows (Fig. 6.4; Brewer *et al.*, 1995), although *P. semisulcatus* also uses algal beds as a nursery habitat (Haywood *et al.*, 1995). It is generally thought that the seagrass beds are highly important for the survival and growth of the juveniles of these commercially valuable prawn species. For other penaeid prawns the requirement of seagrass habitat is less pronounced or even absent: *Penaeus merguiensis*, for instance, is hardly present in seagrass meadows, but is commonly found on bare muddy substrata in mangrove areas; for others, not only seagrasses may serve as a nursery habitat, but also beds of algae, and mud–mangrove banks or channels (Fig. 6.4). The use of alternative nursery habitats next to seagrasses has been established for a number of other decapod crustaceans as well (Sogard & Able, 1991; McMillan *et al.*, 1995).

Spatial and seasonal variability is a common feature of seagrass-associated decapods. Results of a six year study of postlarval and juvenile *Penaeus semisulcatus* in an intertidal mixed seagrass bed (Fig. 6.5) clearly show how abundances change over time, with temporal fluctuations evident at three levels: short-term variation between biweekly samples, seasonal variation on a time scale of months, and annual variation (Vance *et al.*, 1996). Other studies demonstrate considerable variation in

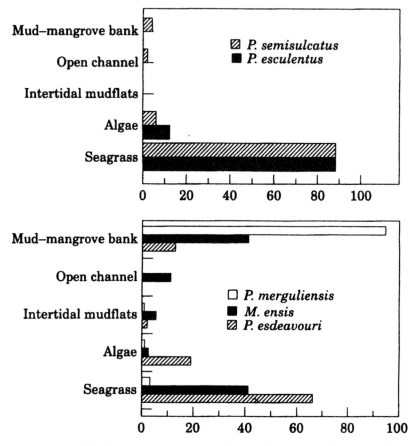

Fig. 6.4. Percentage occurrences of penaeid prawns in different habitat types in the Embley Estuary, Australia. (In Brewer *et al.*, 1995, after data of Staples *et al.*, 1985.)

population densities between meadows in the same region. Some of the factors that were identified as causes of the spatial and seasonal variability in the abundance of fish populations also appear to be responsible for such phenomena in seagrass-associated decapod populations: a locally and temporally different distribution of larvae that are ready to settle, mortality and/or migration following recruitment to the seagrass meadows, and, upon reaching a critical size, the ontogenetic habitat shift from seagrasses to other environments (Gray, 1991; Worthington *et al.*, 1995; Walsh & Mitchell, 1998). Detailed studies of postlarval and juvenile tiger prawns in particular have shed more light on

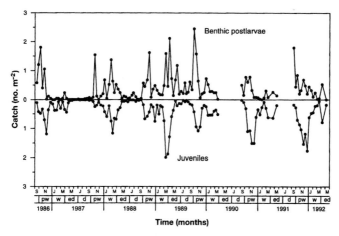

Fig. 6.5. Abundance of postlarval and juvenile *Penaeus semisulcatus* in a seagrass bed in the Embley Estuary, Australia, from Sept. 1986 to May 1992. pw, pre-wet season; w, wet season; ed, early-dry season; d, dry season. (Vance *et al.*, 1996.)

these processes. Small postlarvae of *P. semisulcatus* (with average carapace lengths less than 1.6 mm), offered the choice between the seagrass *Zostera capricorni* and bare sand in laboratory experiments, show no preference for the vegetated habitat, and may even be found in higher numbers over the bare sand habitat. This size class probably largely consists of individuals that are still migrating in the water column, i.e. they appear not to have reached the critical size upon which they become epibenthic. The larger postlarval size classes (with carapace lengths of 1.7 mm or more) demonstrate a change in behaviour: they spend much more time resting on the substrate and perched on seagrass leaves, and show a clear preference for the vegetated habitat during the day (Liu & Loneragan, 1997). This size class of postlarvae is found in the field in shallow seagrass beds. Apparently, at carapace lengths of 1.7 mm individuals become epibenthic, selectively choosing vegetated (structurally complex) habitat for settlement. Interestingly, when large postlarvae and juveniles are offered a choice between *Z. capricorni* (without epiphytes) and artificial seagrass, they show a preference for the living seagrass. This may indicate that a chemical cue released from the seagrass plays a role in the selection of seagrass meadows as a settlement area (Liu & Loneragan, 1997). Lobster postlarvae similarly may select settlement sites in response to chemical agents emanating from their juvenile habitat (Boudreau *et al.*, 1993). *Penaeus semisulcatus* and *P. esculentus* in the Australian Gulf of Carpentaria predominantly settle in intertidal and subtidal seagrass beds (Loneragan *et al.*, 1994). The mechanism that leads to this depth range selectivity is unclear; possibly, the postlarvae that are ready to settle stay close to the water surface, and are transported to the nearshore shallow seagrass beds with incoming tides.

Analysis of the abundance of just settled postlarvae and older individuals in field situations suggests that within days after settlement, the initial pattern of abundance at settlement is modified (Loneragan *et al.*, 1998). This change will be due to either mortality or migration, or a combination of both. Field tests, in which a partitioned container with different types of substratum is deployed in the field to investigate the microhabitat preference of tiger prawns, show that juvenile *Penaeus esculentus* with carapace lengths of 3–10 mm select seagrass with broad, long leaves (*Cymodocea serrulata*) over a species with narrow, long leaves (*Syringodium isoetifolium*), which in turn is preferred over *Halodule uninervis*, a species with narrow, short leaves (Kenyon *et al.*, 1997). Juvenile prawns are active in the water column at night, and are able to cover short distances to reach new habitats. It is thus possible that postsettlement migration and invasion of more preferred seagrass habitats are partly responsible for changes in abundance patterns directly after settlement. The preference of juvenile tiger prawns for broad, long leaves therefore may contribute to the relatively high densities of juveniles with carapace length > 3 mm in natural seagrass beds with the tall, broad-leaved species *Enhalus acoroides* (Loneragan *et al.*, 1998).

The avoidance of predation may underlie the migratory behaviour of juvenile prawns. Predation may also directly, as a proximate factor, contribute to the postsettlement changes in abundance patterns of juveniles in seagrass beds. Other causes, however, cannot be ruled out. Shifts in environmental temperature and salinity towards the limits of physiological tolerance, and competitive interactions between species, have been suggested as direct causes of migration or mortality in caridean shrimps associated with estuarine seagrass meadows (Walsh & Mitchell, 1998).

6.4.3 *Predation in seagrass meadows*

Studies on decapods have contributed their share to the understanding of the mechanistic interactions between predators and prey in seagrass beds. Apart from observations that show that predation of juvenile prawns by fish is lower in seagrass patches than on bare substrate – observations

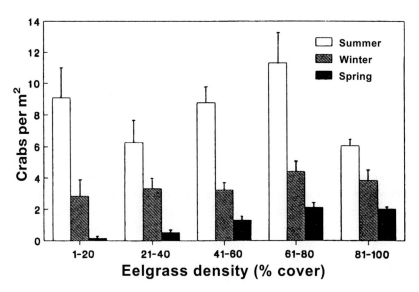

Fig. 6.6. Mean seasonal density (± 1 SE) of Dungeness crab (*Cancer magister*) as a function of *Zostera marina* density in Puget Sound, USA. Combined data of 1984–1987. (McMillan *et al.*, 1995.)

that further corroborate the general notion of seagrasses as a refuge for prey organisms – it appears that the specific morphology of the seagrass can also play a role in predation rates: sand bass (*Psammoperca waigiensis*) is able to detect and consume more juvenile *P. esculentus* in canopies of narrow-leaved *Syringodium isoetifolium* and *Halodule uninervis* than in a canopy of *Cymodocea serrulata*, a species that has relatively long, broad leaves (Kenyon *et al.*, 1995). Juvenile prawns thus can be expected to suffer different predation losses in different types of seagrass meadows, leading to variable decreases in local abundance. As mentioned earlier, shoot density is a prominent element of habitat complexity, and is usually considered to be a major determinant of the degree of refuge provided by seagrass canopies. An example of changes in population abundance after settlement, which are probably related to different, shoot density-dependent predation levels, is provided by a study of dungeness crab (*Cancer magister*) in Puget Sound (USA). Postlarval settlement in *Zostera marina* beds occurs in summer, apparently without regard to density of the plants (Fig. 6.6). The abundance of juveniles, however, changes over time in dependence of eelgrass cover: when spring has come, the eelgrass patches of low shoot density no longer harbour many crabs; the remaining individuals are largely confined to areas with dense vegetation (McMillan *et al.*, 1995). Studies on decapods also provide clear evidence that besides habitat complexity, prey density and species-specific behavioural patterns also matter to predator–prey interactions. These factors may even prevail over habitat complexity in their effect on predation levels. The effectiveness of predators that actively search their prey but are hampered by the disguising plant structures, for instance, is probably reduced by increased habitat complexity, whereas 'ambush' predators that can use the vegetation to reduce their own visibility may profit from it. The lined seahorse (*Hippocampus erectus*) is a predator that may quietly 'sit and wait' for a prey when it can hold on to some object with its prehensile tail. In an environment with seagrass, it may cling to the leaves, waiting for a prey to appear. In laboratory tanks with artificial seagrass, where the seahorse used the 'sit and wait' foraging behaviour, it was found that even the highest seagrass densities (> 3000 blades per m^2) did not reduce the numbers of shrimp that were captured (James & Heck, 1994). Behaviour also appears to influence predation of tiger prawns. Whereas seagrass-associated juvenile *Penaeus esculentus* spends most of its time at night within the bed, *P. semisulcatus* may leave the bed at night and swim in the water column above the seagrasses. This may facilitate predation by pelagic fish, and confound the relationship between habitat complexity and predation levels (Haywood *et al.*, 1998).

6.5 Molluscs

Next to fishes and crustaceans, molluscs can make a conspicuous contribution to the seagrass fauna. An example of the prominent presence of this group is given in Fig. 6.7, which shows the relative abundance of infauna and epifauna in *Zostera marina* beds in the Baltic Sea. Representatives of both gastropods and bivalves are quantitatively important at all localities, and sometimes even dominate the macrofauna present (Boström & Bonsdorff, 1997). The mollusc taxa found in and on the sediment of the eelgrass beds are the same as found outside the beds on bare sand, thereby illustrating the often rather non-specific species composition of seagrass meadows. Numbers, however, are higher in the vegetated sites. The most abundant molluscs in the meadows are small mudsnails (Hydrobiidae), with densities ranging from 3300 to 33 700 ind./m^2 in the *Zostera* beds, and from 300 to 10 900 ind./m^2 on bare sand. Other numerically important molluscs are the bivalve *Macoma balthica* (with 500–6000 ind./m^2 in the *Zostera* beds, and 60–1100 in the

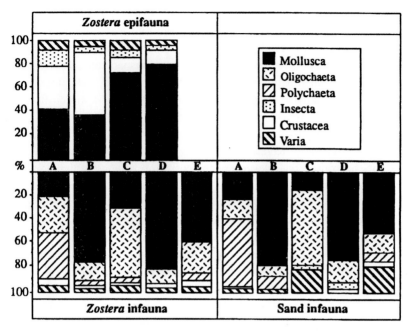

Fig. 6.7. Relative abundance of infauna and epifauna associated with *Zostera marina* beds at 5 localities (A–E) in the northern Baltic Sea. Abundance of infauna of adjacent bare areas is also shown. (Boström & Bonsdorff, 1997.)

unvegetated sites), and another bivalve, the blue mussel *Mytilus edulis* (with 500–6300 ind./m² in the *Zostera* beds, and 60–170 ind./m² in the unvegetated sites). The epifaunal molluscs again were dominated by mudsnails (ca. 800 ind./m²).

Gastropods find various food sources in seagrass beds. There are species that depend on macrophyte tissue for food, and carnivorous gastropods feeding on other members of the faunal community are also present. Most snails, however (including the mudsnails in the *Zostera* beds mentioned above) feed on the abundant microalgae and detritus particles present on the sediment and leaf surfaces. From the perspective of food availability, some bivalves living in seagrass beds would seem to be in a disadvantageous position. This is not so much the case for a bivalve such as *Macoma balthica*: this species has a flexible feeding strategy, and may switch from suspension feeding to deposit feeding when organic particles are available on the sediment surface, reaching out with its long siphon across the sediment to collect them. It is more evident for truly suspension feeding species like *Mytilus edulis*, which collect their food by filtering particles from the water column. Food supply for suspension feeders depends on water flow and water column mixing. It has been regularly observed that suspension feeders grow faster when near-bottom flow velocities are higher, which can be explained by a larger supply of sestonic particles. As water flow is reduced inside seagrass canopies, suspension feeding bivalves thus might experience a reduced food supply in seagrass habitats. The occurrence of suspension feeding bivalve species in the seagrass habitat therefore has raised the question of the costs and benefits of living in this environment.

In the case of the blue mussel, which occurs widely both within and outside seagrass meadows, it has been found that inside *Zostera marina* patches growth may be reduced to only one-third of that in adjacent sand flats (Reusch, 1998). Remarkably, the same study also showed that abundance of 1–5 mm juveniles was considerably higher in eelgrass than in unvegetated areas, indicating successful recruitment to the beds. The positive aspect for mussels of living inside the vegetation may be a decreased chance of dislodgement of the animals during storms. Blue mussels can attach firmly to hard substrates such as rocks with their byssal threads, but on soft substrata they occur as clumps of conspecifics that are just attached to each other. Compared with mussel aggregates on unvegetated bottoms, the risk of being washed ashore by moderate, regularly occurring storms is markedly reduced by the eelgrass canopy, most likely because the canopy reduces current speed and dampens wave energy (Reusch & Chapman, 1995).

Being washed away by storms is less of a risk for burrowing bivalves that live in the sediment, such as the hard clam, *Mercenaria mercenaria*. In unvegetated sediments, growth of this suspension feeding bivalve is faster at higher flow speeds (Grizzle & Morin, 1989). Contrary to expectations based on considerations of reduced flow and food availability inside seagrass canopies, it has been found that growth of *M. mercenaria* actually can be enhanced inside seagrass beds, although this is not consistently the case. Clams may benefit from the seagrass environment in several ways. Bottom-feeding fishes, shrimps and crabs are known to feed on exposed body parts of infaunal invertebrates; the siphons of sediment dwelling bivalves are thus often partially nipped off by epibenthic predators. Clam siphons have been found to be proportionally larger within seagrass beds than on sand flats, suggesting that siphon nipping (with its possibly detrimental effect on growth) is less intense inside the seagrass canopy (Irlandi & Peterson, 1991). Moreover, outside seagrass meadows, clams suffer much higher mortality due to predation by crabs and whelks. Even without direct physical contact, the presence of whelks reduces the time spent feeding by the clam, suggesting that the clams are sensing a chemical signal coming from the whelks. Lower interference with predators within the seagrass canopy, allowing more extended feeding times, thus could contribute to better growth (Irlandi & Peterson, 1991). Larger sediment stability within seagrass beds also plays a beneficial role, particularly for the smaller specimens that have shorter siphons and burrow less deeply in the sediment (Irlandi, 1994, 1996). The lower energetic costs of maintaining a suitable feeding postion in stable sediment relative to more dynamic sedimentary environments may be relevant in that respect.

Studies of the bay scallop *Argopecten irradians* have also given an answer to the question of what benefit the seagrass habitat provides to suspension feeders. Larvae of this commercially exploited species are free-swimming until they settle onto elevated substrates, primarily seagrass leaves. They remain attached for several months, growing rapidly. Because the leaves are only an ephemeral substrate, the scallops continuously have to re-attach to new leaves to maintain their elevated position in the canopy. After this first phase in the seagrass bed, the scallops move to the bottom, where they live unattached in between the shoots. Field experiments with tethered scallops of 10–15 mm size in *Zostera marina* meadows showed that survival increased at greater heights of attachment: at 20–35 cm above the bottom, over 59% of the juvenile scallops had survived after 4 days, whereas near the sediment surface survival was less than 11%. Mortality was mostly due to predation by several crab species and starfish. Settlement and attachment on seagrass leaves thus is an effective means of avoiding high mortalities from predators that forage primarily on the bottom. With growth, predation risks decrease, and at sizes from 30–40 mm onwards, when scallops have relocated to the bottom, the risk of falling prey to benthic predators is relatively small. The scallops thus experience a shift from a spatial to a size refuge from predation over the course of their postsettlement life in the eelgrass habitat (Pohle *et al.*, 1991). The selection by juveniles of attachment sites high in the canopy may be at the expense of growth rates. Experiments with 12 mm juveniles showed that while survival on the bottom was significantly lower than at 15 cm elevation in an artificial seagrass canopy, scallops grew fastest on the bottom. These observations thus suggest that juvenile scallops are faced with a trade-off between minimizing their risk of predation and maximizing their growth (Ambrose & Irlandi, 1992).

Considering the various results that have been obtained with *Mytilus edulis*, *Mercenaria mercenaria* and *Argopecten irradians*, it is clear that generalizations on seagrass–bivalve interactions cannot yet be readily made. Growth of bivalves may be both positively and negatively affected in seagrass meadows, and benefits in terms of survival appear to be derived from processes that are different for different species, and that vary between sites. It should be added here that the notion of a lower food availability for suspension feeding bivalves in seagrass meadows is not unchallenged. Evidence has been presented that *Mercenaria mercenaria* may encounter more concentrated (locally derived, authochthonous) microalgal food in the water layer just above the sediment surface in the seagrass bed than in the water layer above bare sand habitats (Judge *et al.*, 1993). A higher food supply may also be found in meadows of *Posidonia oceanica*. These meadows are the habitat of the fan mussel, *Pinna nobilis* (Fig. 6.8). This bivalve reaches lengths of 86 cm (Moreteau & Vicente, 1982), protruding for about 80% of its length above the sediment surface. Hence, it is only hidden from predators within the high canopy (up to 100 cm) of the seagrass. Despite the fact that the animal lives in an oligotrophic sea, it displays the fastest growth rate yet reported for any bivalve (up to 1 mm per day; Richardson *et al.*, 1999), which can only be accounted for an the enhanced food availability present within the canopies of *Posidonia oceanica* (Duarte *et al.*, 1999). Further studies on the actual food supply for suspension feeding bivalves inside seagrass canopies are thus warranted.

6.6 Endangered grazers: turtles and sea cows

The largest animals that are associated with the seagrass habitat are the green turtle, *Chelonia mydas*, and species of the order Sirenia (sea cows), notably the dugong *Dugong dugon*, and the West Indian manatee *Trichechus manatus*. These animals are herbivores, and forage over seagrass meadows. A second manatee species, *T. senegalensis* (the West African manatee) may also consume seagrass, but data on this animal are scanty. Considerable conservation efforts are now devoted to prevent extinction of these spectacular animals.

Fig. 6.8. *Pinna nobilis*, about 60 cm tall, with its densely epiphytized shell growing in a *Posidonia oceanica* meadow. (Photograph by C.M. Duarte.)

6.6.1 Turtles

The feeding habits of some of the seven species of sea turtles are still poorly known, but the observations so far indicate that the dependence of *Chelonia mydas* on macrophyte food is unique. All other species are probably mainly carnivorous and consume primarily benthic and pelagic invertebrates, such as crabs, molluscs, sponges and jellyfish. Stomach content analyses indicate that some species, notably the hawksbill (*Eretmochelys imbricata*), may ingest seagrass and algae as well, but the full spectrum of observations indicates that these food sources are of minor importance (Bjorndal, 1997).

After leaving their nesting beaches, *Chelonia mydas* spends the first years of its life in the open ocean. During this life stage, it is probably omnivorous. When the turtles have reached a size of 20–35 cm, they leave the pelagic habitat and enter benthic foraging areas. This habitat shift coincides with a shift to a primarily herbivorous diet, although the animals are not averse to eating animal matter (Bjorndal, 1997). Sea-grasses and algae, however, are the main sources of food. The green turtle has a wide distribution in tropical and subtropical waters, and occurs in the Atlantic, Pacific and Indian oceans. The specific macro-phyte species composition of the diet varies with the geographic location. Throughout most of their distributional area the turtles feed primarily on seagrasses, algae being the principal food where seagrass meadows are lacking (Mortimer, 1995). Mixed diets of both seagrasses and algae are also not uncommon (Lanyon *et al.*, 1989). As far as the consumption of seagrasses is concerned, the turtles are not very fastidious. A number of species have been recorded as food plants, belonging to different genera: *Thalassia testudinum* (turtle grass), *Syringodium* spp., *Halophila* spp., *Halodule* spp., *Cymodocea serrulata*, *Thalassodendron ciliatum*, *Posidonia oceanica* and *Zostera* spp. (Mortimer, 1995; Bjorndal, 1997). The green turtle is nowadays a threatened species, due to a variety of anthropogenic causes including the overexploitation of local populations, unintentional capture in fish nets, and habitat destruction. In earlier times, however, populations may have been orders of magnitude greater, making the species a major seagrass consumer in its distribution range (Bjorndal, 1980).

The observations on feeding behaviour in seagrass beds indicate that turtles may display a diel foraging pattern. They rest during the night (e.g. in deep holes or near sheltering reef patches), and swim to the seagrass beds in the morning to forage during the day time (Bjorndal,

1980; Williams, 1988). Observations on ingested seagrass material indicate that the turtles selectively consume younger leaves and leaf parts; the epiphytized older leaves and leaf parts are avoided (Mortimer, 1981; Lanyon *et al.*, 1989). The green turtle thus apparently does not exploit the epiphytic algae and sessile epifauna that usually cover older leaf surfaces as food sources, but fully relies on seagrass tissue. With age, seagrass leaf tissues decrease in nutrient content (Stapel & Hemminga, 1997). Young, growing leaves also will have relatively low contents of lignin, a cell wall substance that deters herbivores. The preference of green turtles for palatable, young leaf tissue probably explains the observation that the animals repeatedly may return to the same, discrete areas within a seagrass vegetation. Within these grazing plots all leaves are clipped off to a height of a few centimetres, and the animals continually revisit these plots to harvest the young, newly formed leaves (Bjorndal, 1980). Repeated leaf cropping stresses the plants and ultimately results in a decline of leaf production, which has been ascribed to a depletion of the sediment nutrients available for plant growth (Zieman *et al.*, 1984). The decline in new leaf formation probably is the reason that the turtles abandon the grazing plots after some time (this may be as long as a year or even more), to create new ones elsewhere. Repeated grazing of discrete areas is not a rigid feeding strategy in green turtles, and the common occurrence of this behaviour remains to be confirmed. When the seagrass vegetation has a low density, is scattered, or has a poor forage quality, the turtles probably tend to move over larger distances and in a less consistent pattern through their foraging area, in order to meet their nutritional needs (Williams, 1988; Whiting & Miller, 1998). Repeated grazing has also not been observed in juvenile turtles (Ogden *et al.*, 1983), and thus may be restricted to older animals.

6.6.2 Sea cows

The distribution area of the West Indian manatee comprises the coastal and island waters from Florida, the Gulf of Mexico and the Caribbean, extending southward along the coasts of South America to north Brazil. The dugong lives in the tropical and subtropical waters of the Indian and western Pacific oceans. Although both species may deliberately ingest invertebrates (Powell, 1978; Preen, 1995a), they are essentially herbivores. The West Indian manatee moves between fresh and salt water, and its diet consequently comprises a variety of both freshwater and marine food plants. Hence, although it may feed on seagrasses (Hartman, 1979), it is not a true seagrass specialist. In contrast, the dugong is strictly marine and feeds nearly exclusively on seagrasses. This rather slow-moving species, which must venture into shallow coastal waters in order to obtain its food, is an easy target, and hunting pressures in particular have diminished the abundance of these animals in past centuries. However, fairly large numbers still survive, particularly in Australian waters, in the Arabian Gulf and in the Red Sea. The vast coastal waters of Australia host by far the largest remaining populations. At the end of the eighties the number of dugongs in these waters was estimated at some tens of thousands (Lanyon *et al.*, 1989).

Dugongs are known to consume different seagrass species: *Amphibolis antarctica*, *Cymodocea* spp., *Halodule* spp., *Halophila* spp., *Thalassia hemprichii*, *Enhalus acoroides*, *Syringodium isoetifolium*, *Thalassodendron ciliatum* and *Zostera capricorni* have been recorded as food plants (Lanyon *et al.*, 1989). Field observations and observations on captive animals, however, indicate that dugongs have a preference for species of the genera *Halophila* and *Halodule* (Anderson, 1994; Preen, 1995b; De Iongh, 1996; De Iongh *et al.*, 1995). A remarkable aspect of feeding by the dugong is that the entire plant, including the roots and rhizomes, may be harvested, a feature that distinguishes the sea cow from the green turtle, which only feeds on leaves. Feeding on entire plants does not always occur, but seems to depend on the structural characteristics of the seagrass species. Dugongs feeding on *Amphibolis antarctica*, *Enhalus acoroides* or *Thalassia hemprichii* only browse on the leaf clusters, and seem to be unable to harvest the tough rhizomes (Marsh *et al.*, 1982; Anderson, 1986; Erftemeijer, 1993). The more delicate belowground structures of *Halophila* spp. and *Halodule* spp., however, present no problem and are efficiently cropped, together with the leaf clusters. By consuming the rhizomes, the dugongs are able to increase their energy intake, as they get access to the rich carbohydrate reserves that are stored in these organs. Being able to harvest the rhizome tissues is probably the major reason underlying the dietary preference for *Halophila* and *Halodule* (Anderson, 1994, 1998; De Iongh *et al.*, 1995).

Dugongs have the habit of uprooting and consuming the plants in an uninterrupted, metres-long trail through the seagrass. This results in characteristic, meandering bare feeding trails, which are about as wide as their muzzles, and in which both the aboveground and the belowground plant biomass has largely been removed (Preen, 1995b; De Iongh *et al.*, 1995). An example of the cumulative pattern of feeding trails observed in a seagrass-covered bay in the Haruku Strait (Indonesia), over a 3-month

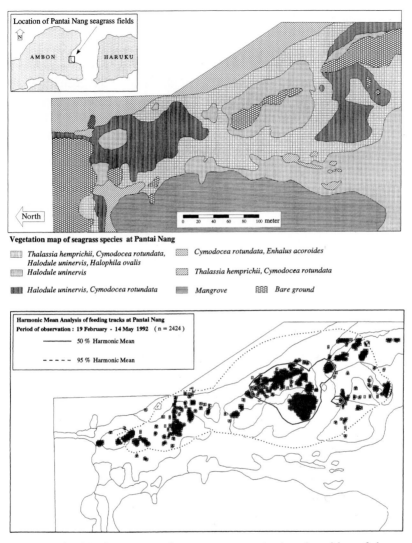

Fig. 6.9. Seagrass species-coverage map (*top*) and position of dugong feeding tracks observed from February to May 1992 (*bottom*) in an embayment of Haruku Strait, Indonesia. (De Iongh, 1996.)

period, is shown in Fig. 6.9. The seagrass cover in this bay consists of 45% of *Halodule uninervis* dominated beds, while mixed vegetation of *Cymodocea rotundata*, *Thalassia hemprichii* and *Enhalus acoroides* covers the remaining surface. The feeding trails are clearly concentrated in the *H. uninervis* dominated vegetation which illustrates the preference of the dugongs for this pioneer species. Monitoring of new feeding trails during the period of observation, furthermore, indicated that the animals repeatedly returned to the same restricted foraging areas (De Iongh, 1996). This type of grazing, i.e. concentrated at a particular location and maintained over an extended period, is similar to what has been observed in Australian waters, where large herds of more than 100 dugongs often graze at the same location for weeks to months (Preen, 1993). Experiments in which dugong grazing is simulated show that the seagrass species composition can be altered by intensive grazing. The rapidly growing pioneer species *Halophila ovalis* (the first to recolonize the bare feeding trails) is favoured, at the expense of slower growing species (Preen, 1995b). The concentrated grazing of dugongs at the same location and the repeated cropping of the vegetation thus may benefit the dugongs because it furthers the creation and maintenance of beds of its preferred food plants, *Halophila* and *Halodule* pioneer species (Preen, 1995b; De Iongh, 1996).

The virtual complete dependence of dugongs on seagrasses as a food source makes them very vulnerable to losses of seagrass habitat. This is evident from the population decline in Hervey Bay (Australia) after an unusual combination of floods and a cyclone had led to the large-scale disappearance of seagrass vegetation (1000 km^2) in the bay (Preen *et al.*, 1995). From the initial population of about 2000 animals, only some 70 were left 8 months after the floods. Many dugong carcasses were found in the months following the loss of the foraging grounds, frequently showing clear starvation symptoms (Preen & Marsh, 1995).

6.7 Faunal activity and seagrass growth and production

This chapter so far has primarily focused on the role of seagrasses as a habitat and foraging area for animals. The seagrass fauna, however, also plays a role in the functioning of the seagrasses. In this section, we will address some aspects of the impact of the fauna on seagrass growth and production. Three types of effects can be distinguished: (1) direct effects of grazing on seagrass, i.e. the effects of tissue removal on the macrophyte vegetation, (2) indirect effects of feeding activities, i.e. effects of periphyton grazing and impacts on nutrient cycling, and, (3) physical disturbances of the seagrass habitat, associated with the presence or bioturbation activity of some animals.

6.7.1 Direct effects of grazing on seagrass

As discussed in sections 5.3 and 6.6, direct grazing on seagrasses may have a severe impact on the vegetation. Studies of grazing by sea urchins during periods of very high population density, of the periodical consumption of aboveground and belowground tissues by large migratory flocks of waterfowl, and of the concentrated and continued feeding activity of turtles and sea cows, provide examples of how herbivorous animals can locally drastically decrease seagrass biomass, reduce productivity, or induce shifts in species composition. At moderate grazing pressures, negative impacts on leaf growth are probably generally small (Cebrián *et al.*, 1998). In terrestrial grasslands, herbivores have actually been shown to enhance plant production. A recent experimental study of the effects of sea urchin grazing on *Thalassia testudinum* in the Gulf of Mexico indicates that during the growing season, turtle grass responds to grazing by increasing the recruitment of shoots. This results in an increase in net aboveground production, and thus resembles the findings in terrestrial grasslands (Valentine *et al.*, 1997). Results obtained by Cebrián *et al.* (1998), however, suggest that such a compensatory growth response is not a general phenomenon. Furthermore, it may also be transient and disappear as harvesting continues (Zieman *et al.*, 1984).

In some instances, grazing may have deleterious effects on seagrass that are disproportional relative to the amounts of tissue that are ingested by the herbivores. The bites of herbivores may lead to mechanical weakening or the complete loss of the leaf–shoot connection resulting in detachment of large leaf parts; this is a commonly observed collateral effect of herbivorous activity. In *Zostera marina* it has been found that grazing of the leaf epidermis by the small commensal limpet *Tectura depicta* has a devastating effect on the carbon balance of the plant, the reason being that the epidermis, although it represents a minor fraction of the total leaf biomass, contains virtually all of the leaf chlorophyll; consumption of this epidermis thus directly affects the plant's capacity for photosynthetic carbon fixation (Zimmerman *et al.*, 1996). Even the feeding activity of minute isopod miners, observed in the leaves of *Posidonia australis* and *P. sinuosa*, may have consequences beyond the direct consumption losses as burrows facilitate entry of water and microorganisms that may provoke localized tissue death (Brearly & Walker, 1995).

6.7.2 Indirect effects of feeding activities

A variety of animals feeds on the periphyton layer, with isopods, amphipods and gastropods being the taxa that commonly play the most prominent role. Some of these animals feed quite selectively, targeting on a limited set of microalgal species; others ingest the periphyton layer without much preference for any of its components. In each case, the effect is the partial removal of a layer that potentially intercepts light before it reaches the seagrass leaf epithelium, and interferes with the supply of inorganic nutrients and carbon. Another positive effect of periphyton grazers could be a reduction of the risk of leaf loss or leaf fragmentation due to hydrodynamical forces: heavy epiphyte loads may reduce the flexibility of the leaves and increase drag, leading to breaking and severance of these plant parts (Jernakoff *et al.*, 1996). Scanning electron micrographs give a vivid picture of the 'cleaning' effect of grazers (Fig. 6.10). Particularly because light limitation plays such a crucial role in the productivity and vertical distribution of seagrasses, the extent to which epifaunal grazers influence seagrass growth by removal of the periphyton or epiphyte layer has received considerable attention. For extensive reviews of the subject the reader is referred to reviews by Van Montfrans *et al.* (1984), and Jernakoff *et al.* (1996). A general outcome of the various studies on periphyton–grazer interactions is that grazers reduce periphyton biomass on seagrass leaves in comparison with ungrazed leaves. The effect of grazing on periphyton biomass, however, may differ quantitatively and qualitatively depending on the epifaunal species involved. At natural densities, the isopod *Idotea chelipes*, for instance, appears to be much more effective in the removal of periphyton biomass from *Zostera marina* leaves than the snail *Littorina littorea* (Hootsmans & Vermaat, 1985). In another study, with exclosure chambers harbouring *Posidonia sinuosa* with either an amphipod assemblage or the herbivorous snail *Thalotia conica*, it was found that the gastropod reduced the biomass of larger epiphytic algae, but had no effect on the number of algal taxa, whereas the amphipods only temporarily reduced the biomass of the algae, but increased the number of taxa by 29%. Differences in food selectivity, possibly related to the functional morphology of the mouthparts, were held responsible for the different impacts of these grazers, the amphipods being more specific in their food choice than the snail species (Jernakoff & Nielsen, 1997).

The performance of seagrass vegetation with grazers has been repeatedly compared with that of seagrass in the absence of grazers. In several

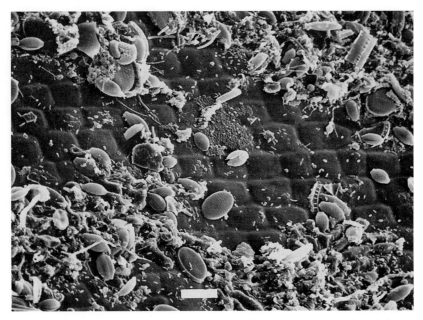

Fig. 6.10. Grazing trail of the snail *Bittium varium* through periphyton on *Zostera marina*. In the grazing trail the cobblestone-like leaf epithelium is visible as the periphyton layer is nearly completely removed. Bar: 10 μm. (Photograph obtained by courtesy of R.J. Orth.)

studies it has been found that removal of periphyton by grazers enhances the productivity and biomass of seagrasses (Hootsmans & Vermaat, 1985; Howard & Short, 1986; Philippart, 1995). Seagrass growth thus seems to be partly under indirect control of grazers, with grazers mitigating the negative impact of the periphyton. This may be particularly important when ambient nutrient levels are high. Overgrowth of epiphytes at elevated water column nutrient levels has been pointed out as one of the mechanisms that may result in large-scale declines of seagrass vegetation (see section 7.5.1). It is thought that the presence of grazers can play a crucial role in keeping the epiphyte load on the leaves in check, thereby maintaining the vitality of the macrophytes. Experimental observations, showing that an increase in epiphyte biomass during water column fertilization can be prevented by effective epiphyte herbivores, support this notion (Williams & Ruckelshaus, 1993). Studies of the relationships between nutrient levels, epiphyte grazers and seagrass growth, however, also indicate that these relationships are not simple or consistent. The effect of grazers on epiphyte loads and, indirectly, on seagrasses, varies seasonally, depending on grazing pressures and abiotic controls on seagrass and epiphyte growth (Neckles *et al.*, 1993).

Indirect feeding effects of fauna also include impacts on nutrient dynamics in the meadows. Each animal species living in the seagrasses contributes, by its feeding and metabolic activities, to the cycling of nutrients on which growth of the seagrasses ultimately depends. In the case of the blue mussels, *Mytilus edulis*, the effect on nutrient fluxes and, concomitantly, on seagrass growth, has been investigated in some detail. At moderate densities (less than 20% areal coverage), the presence of clumps of *Mytilus edulis* in eelgrass beds has been found to enhance the size of the eelgrass leaves. This effect is mediated by the continuous production of nutrient-rich faeces and pseudofaeces by these suspension feeding bivalves. The mineralization of the organic material on the sediment surface results in the release of inorganic nutrients and in an elevation of the porewater concentration of ammonium and phosphate. In this way, the sediment is fertilized and plant growth is stimulated in nutrient-limiting situations (Reusch *et al.*, 1994).

6.7.3 *Effects of physical disturbances*

The third type of faunal effects on seagrass growth and production is by physical disturbance of the habitat. Churning up the rooted sediment can have negative consequences for seagrass growth and distribution. The burrowing and mound building activities of larger shrimps and crabs are examples of this type of disturbance (Suchanek, 1983; Valentine *et al.*, 1994; Woods & Schiel, 1997). Bivalves may also induce a shift of the physical characteristics of the seagrass environment. The mussel *Musculista senhousia* forms a cocoon of byssus threads and sediment particles. At high densities these cocoons fuse to form a more or less continuous carpet in the topmost sediment layer. Spatial interference of this tough byssal carpet with roots and rhizomes are thought to be the cause of diminished rhizome elongation rates in *Zostera marina* meadows that harbour the bivalve (Reusch & Williams, 1998). At high densities, such as may occur following heavy spatfall, blue mussels (*Mytilus edulis*) may also have a negative effect on eelgrass meadows. The clumps of mussels lying on top of the sediment can be so compact that plant growth is physically impeded, to the extent that the seagrass vegetation completely disappears, to be displaced by a dense mussel bank (Gründel, 1980, cited in Reusch *et al.*, 1994).

6.8 Concluding remarks

Many animals are permanent or temporary residents of seagrass meadows. Major factors that attract fauna are the protection that the seagrass canopy offers from predation, and the presence of food. As discussed in the previous section, the presence of these animals is not without consequences for the functioning of the seagrass vegetation. Feeding activities have direct and indirect effects on the vegetation, and animals may also influence the seagrasses by more or less profound physical disturbance of the environment. The abundance and richness of seagrass fauna have been stressed as important and characteristic attributes of seagrass systems by numerous authors, undoubtedly in part also as an argument for the protection of these rather inconspicuous systems, which lack the glamour of other coastal systems such as coral reefs and mangroves. Indeed, comparison of data clearly shows that the abundance and diversity of the fauna that is associated with seagrass meadows is high compared with that of unvegetated areas. It is also a fact that seagrass systems can have considerable economical value as nursery habitats for commercially important animal species; in addition, they are a nursery habitat for numerous non-commercial species. For some species, seagrass meadows are essential for their survival, and this includes those unique marine mammals, the dugongs. A balanced picture of the fauna of seagrass meadows, however, requires some additional remarks. The wealth of the literature on fauna in seagrass systems indicates that the faunal assemblages of seagrass meadows are not specific, but largely comprise species occurring outside the seagrass meadows as well. Furthermore, other vegetated systems in the direct environment of seagrass meadows may harbour an equally abundant and diverse fauna (or an even richer one), indicating that seagrass meadows offer no exceptionally favourable habitat compared with other systems. Finally, with respect to their nursery value, it is evident that not *all* seagrass meadows are equally valuable as nursery areas. Seagrasses have been said to provide shallow marine areas with a 'valuable benthic substratum' (Howard *et al.*, 1989), and this modest description seems to characterize the significance of seagrasses as a habitat for fauna, as it emerges from the literature, particularly well.

6.9 References

Ambrose, W.G., Jr. & Irlandi, E.A. (1992). Height of attachment on seagrass leads to trade-off between growth and survival in the bay scallop *Argopecten irradians*. *Marine Ecology Progress Series*, **90**, 45–51.

Anderson, P.K. (1986). Dugongs of Shark Bay, Australia – seasonal migration, water temperature and forage. *National Geographic Research*, **2**, 473–90.

Anderson, P.K. (1994). Dugong distribution, the seagrass *Halophila spinulosa*, and thermal environment in winter in deeper waters of Eastern Shark Bay, Western Australia. *Wildlife Research*, **21**, 381–8.

Anderson, P.K. (1998). Shark Bay dugongs (*Dugon dugong*) in summer: II: foragers in a *Halodule*-dominated community. *Mammalia*, **62**, 409–25.

Ansari, Z.A., Rivonker, C.U., Ramani, P. & Parulekar, A.H. (1991). Seagrass habitat complexity and macroinvertebrate abundance in Lakshadweep coral reef lagoons, Arabian Sea. *Coral Reefs*, **10**, 127–31.

Baelde, P. (1990). Differences in the structures of fish assemblages in *Thalassia testudinum* beds in Guadeloupe, French West Indies, and their ecological significance. *Marine Biology*, **105**, 163–73.

Bell, J.D. & Westoby, M. (1986a). Abundance of macrofauna in dense seagrass is due to habitat preference, not predation. *Oecologia*, **68**, 205–9.

Bell, J.D. & Westoby, M. (1986b). Importance of local changes in leaf height and density to fish and decapods associated with seagrasses. *Journal of Experimental Marine Biology and Ecology*, **104**, 249–74.

Bell, J.D. & Westoby, M. (1986c). Variation in seagrass height and density over a wide spatial scale: effects on common fish and decapods. *Journal of Experimental Marine Biology and Ecology*, **104**, 275–95.

Bjorndal, K.A. (1980). Nutrition and grazing behavior of the green turtle *Chelonia mydas*. *Marine Biology*, **56**, 147–54.

Bjorndal, K.A. (1997). Foraging ecology and nutrition of sea turtles. In *The Biology of Sea Turtles*, ed. P.L. Lutz & J.A. Musick, pp. 199–231. CRC Press.

Blaber, S.J.M., Brewer, D.T., Salini, J.P., Kerr, J.D. & Conacher, C. (1992). Species composition and biomasses of fishes in tropical seagrasses at Groote Eylandt, Northern Australia. *Estuarine, Coastal and Shelf Science*, **35**, 605–20.

Boström, C. & Bonsdorff, E. (1997). Community structure and spatial variation of benthic invertebrates associated with *Zostera marina* (L.) beds in the northern Baltic Sea. *Journal of Sea Research*, **37**, 153–66.

Boudreau, B., Bourget, E. & Simard, Y. (1993). Behavioural responses of competent lobster postlarvae to odor plumes. *Marine Biology*, **117**, 63–9.

Brearley, A. & Walker, D.I. (1995). Isopod miners in the leaves of two Western Australian *Posidonia* species. *Aquatic Botany*, **52**, 163–81.

Brewer, D.T., Blaber, S.J.M., Salini, J.P. & Farmer, M.J. (1995). Feeding ecology of predatory fishes from Groote Eylandt in the Gulf of Carpentaria, Australia, with special reference to predation on Penaeid prawns. *Estuarine, Coastal and Shelf Science*, **40**, 577–600.

Burchmore, J.J., Pollard, D.A., Middleton, M.J., Bell, J.D. & Pease, B.C. (1988). Biology of four species of whiting (Pisces: Sillaginidae) in Botany Bay, New South Wales. *Australian Journal of Marine and Freshwater Research*, **39**, 709–27.

Cebrián, J., Duarte, C.M., Agawin, N.S.R. & Merino, M. (1998). Leaf growth response to simulated herbivory: a comparison among seagrass species. *Journal of Experimental Marine Biology and Ecology*, **220**, 67–81.

Coles, R.G., Lee Long, W.J., Watson, R.A. (1993). Distribution of seagrasses, and their fish and Penaeid prawn communities, in Cairns Harbour, a tropical estuary, Northern Queensland, Australia. *Australian Journal of Marine and Freshwater Research*, **44**, 193–210.

Connolly, R.M. (1994a). A comparison of fish assemblages from seagrass and unvegetated areas of a southern Australian estuary. *Australian Journal of Marine and Freshwater Research*, **45**, 1033–44.

Connolly, R.M. (1994b). Removal of seagrass canopy: effects on small fish and their prey. *Journal of Experimental Marine Biology and Ecology*, **184**, 99–110.

Connolly, R.M. (1997). Differences in composition of small, motile invertebrate assemblages from seagrass and unvegetated habitats in a southern Australian estuary. *Hydrobiologia*, **346**, 137–48.

De Iongh, H.H. (1996). Plant-herbivore interactions between seagrasses and dugongs in a tropical small island ecosystem. Ph.D. Thesis, Catholic University Nijmegen, The Netherlands.

De Iongh, H.H., Wenno, B.J., & Meelis, E. (1995). Seagrass distribution and seasonal changes in relation to dugong grazing in the Moluccas, East Indonesia. *Aquatic Botany*, **50**, 1–19.

Duarte, C.M., Benavent, E. & Sánchez, M.C. (1999). The microcosm of particles within seagrass (*Posidonia oceanica*) canopies. *Marine Ecology Progress Series*, **181**, 289–95.

Edgar, G.J. (1990). Population regulation, population dynamics and competition amongst mobile epifauna associated with seagrass. *Journal of Experimental Marine Biology and Ecology*, **144**, 205–34.

Edgar, G.J. & Shaw, C. (1995a). The production and trophic ecology of shallow-water fish assemblages in southern Australia I. Species richness, size-structure and production of fishes in Western Port, Victoria. *Journal of Experimental Marine Biology and Ecology*, **194**, 53–81.

Edgar, G.J. & Shaw, C. (1995b). The production and trophic ecology of shallow-water fish assemblages in southern Australia II. Diets of fishes and trophic relationships between fishes and benthos at Western Port, Australia. *Journal of Experimental Marine Biology and Ecology*, **194**, 83–106.

Edgar, G.J. & Shaw, C. (1995c). The production and trophic ecology of shallow-water fish assemblages in southern Australia III. General relationships between sediments, seagrasses, invertebrates and fishes, *Journal of Experimental Marine Biology and Ecology*, **194**, 107–31.

Edgar, G.J., Shaw, C., Watson, G.F. & Hammond, L.S. (1994). Comparisons of species richness, size-structure and production of benthos in vegetated and unvegetated habitats in Western Port, Victoria. *Journal of Experimental Marine Biology and Ecology*, **176**, 201–26.

Erftemeijer, P.L.A., Djunarlin & Moka, W. (1993). Stomach content analysis of a dugong (*Dugong dugon*) from South Sulawesi, Indonesia. *Australian Journal of Marine and Freshwater Research*, **44**, 229–33.

Ferrell, D.J., McNeill, S.E.M., Worthington, D.G. & Bell, J.D. (1993). Temporal and spatial variation in the abundance of fish associated with the seagrass *Posidonia australis* in South-eastern Australia. *Australian Journal of Marine and Freshwater Research*, **44**, 881–99.

Francour, P. (1997). Fish assemblages of *Posidonia oceanica* beds at Port-Cros (France, NW Mediterranean): assessment of composition and long-term fluctuations by visual census. *P.S.Z.N. I: Marine Ecology*, **18**, 157–73.

242   *Fauna associated with seagrass systems*

Gambi, M.C., Conti, G. & Bremec, C.S. (1998). Polychaete distribution, diversity and seasonality related to seagrass cover in shallow soft bottoms of the Tyrrhenian Sea (Italy). *Scientia Marina*, **62**, 1–17.

Gotceitas, V. & Colgan, P. (1989). Predator foraging success and habitat complexity: quantitative test of the threshold hypothesis. *Oecologia*, **80**, 158–66.

Gotceitas, V., Fraser, S. & Brown, J.A. (1997). Use of eelgrass beds (*Zostera marina*) by juvenile Atlantic cod (*Gadus morhua*). *Canadian Journal of Fisheries and Aquatic Sciences*, **54**, 1306–19.

Gray, C.A. (1991). Demographic patterns of the palaemonid prawn *Macrobrachium intermedium* in southeastern Australia: spatial heterogeneity and the effects of species of seagrass. *Marine Ecology Progress Series*, **75**, 239–49.

Gray, C.A., McElligott, D.J. & Chick, R.C. (1996). Intra- and inter-estuary differences in assemblages of fishes associated with shallow seagrass and bare sand. *Marine and Freshwater Research*, **47**, 723–35.

Gray, C.A., Chick, R.C. & McElligott, D.J. (1998). Diel changes in assemblages of fishes associated with shallow seagrass and bare sand. *Estuarine, Coastal and Shelf Science*, **46**, 849–59.

Grizzle, R.E. & Morin, P.J. (1989). Effect of tidal currents, seston, and bottom sediments on growth of *Mercenaria mercenaria*: results of a field experiment. *Marine Biology*, **102**, 85–93.

Gründel, E.R. (1980) (cited in Reusch *et al.*, 1994). Ökosystem Seegraswiese: qualitative und quantitative Untersuchungen über Struktur und Funktion einer *Zostera*-Wiese vor Surendorf (Kieler Bucht, Westliche Ostsee). Ph.D. Thesis, University of Kiel.

Hamer, P.A. & Jenkins, G.P. (1996). Larval supply and short-term recruitment of a temperate zone demersal fish, the King George whiting, *Sillaginodes punctata* Cuvier and Valenciennés, to an embayment in south-eastern Australia. *Journal of Experimental Marine Biology and Ecology*, **208**, 197–214.

Hannan, J.C. & Williams, R.J. (1998). Recruitment of juvenile marine fishes to seagrass habitat in a temperate Australian estuary. *Estuaries*, **21**, 29–51.

Hartman, D.S. (1979). Ecology and behavior of the manatee (*Trichechus manatus*) in Florida. Special publication of the American Society of Mammologists, **5**, 1–153.

Haywood, M.D.E., Vance, D.J. & Loneragan, N.R. (1995). Seagrass and algal beds as nursery habitats for tiger prawns (*Penaeus semisulcatus* and *P. esculentus*) in a tropical Australian estuary. *Marine Biology*, **122**, 213–23.

Haywood, M.D.E., Heales, D.S., Kenyon, R.A., Loneragan, N.R. & Vance, D.J. (1998). Predation of juvenile tiger prawns in a tropical Australian estuary. *Marine Ecology Progress Series*, **162**, 201–14.

Heck, K.L., Jr., Able, K.W., Fahay, M.P. & Roman, C.T. (1989). Fishes and decapod crustaceans of Cape Cod eelgrass meadows: species composition, seasonal abundance patterns and comparison with unvegetated substrates. *Estuaries*, **12**, 59–65.

Heck, K.L. Jr., Able, K.W., Roman, C.T. & Fahay, M.P. (1995). Composition, abundance, biomass, and production of macrofauna in a New England estuary: comparisons among eelgrass meadows and other nursery habitats. *Estuaries*, **18**, 379–89.

Hootsmans, M.J.M. & Vermaat, J.E. (1985). The effect of periphyton-grazing by three epifaunal species on the growth of *Zostera marina* L. under experimental conditions. *Aquatic Botany*, **22**, 83–8.

Howard, R.K. & Short, F.T. (1986). Seagrass growth and survivorship under the influence of epiphyte grazers. *Aquatic Botany*, **24**, 287–302.

Howard, R.K., Edgar, G.J. & Hutchings, P.A. (1989). Faunal assemblages of seagrass beds. In *Biology of Seagrasses*, ed. A.W.D Larkum, A.J. McComb & S.A. Shepherd, pp. 536–64. Amsterdam: Elsevier.

Hyndes, G.A., Potter, I.C. & Lenanton, R.C.J. (1996). Habitat partitioning by whiting species (Sillaginidae) in coastal waters. *Environmental Biology of Fishes*, **45**, 21–40.

Irlandi, E.A. (1994). Large- and small-scale effects of habitat structure on rates of predation: how percent coverage of seagrass affects rates of predation and siphon nipping on an infaunal bivalve. *Oecologia*, **98**, 176–83.

Irlandi, E.A. (1996). The effects of seagrass patch size and energy regime on growth of a suspension-feeding bivalve. *Journal of Marine Research*, **54**, 161–85.

Irlandi, E.A. & Peterson, C.H. (1991). Modification of animal habitat by large plants: mechanisms by which seagrasses influence clam growth. *Oecologia*, **87**, 307–18.

James, P.L. & Heck, K.L., Jr. (1994). The effects of habitat complexity and light intensity on ambush predation within a simulated seagrass habitat. *Journal of Experimental Marine Biology and Ecology*, **176**, 187–200.

Jenkins, G.P. & Wheatley, M.J. (1998). The influence of habitat structure on nearshore fish assemblages in a southern Australian embayment: comparison of shallow seagrass, reef-algal and unvegetated sand habitats, with emphasis on their importance to recruitment. *Journal of Experimental Marine Biology and Ecology*, **221**, 147–72.

Jenkins, G.P., Wheatley, M.J. & Poore, A.G.B. (1996). Spatial variation in recruitment, growth, and feeding of postsettlement King George whiting, *Sillaginodes punctata*, associated with seagrass beds of Port Phillip Bay, Australia. *Canadian Journal of Fisheries and Aquatic Sciences*, **53**, 350–9.

Jenkins, G.P., May, H.M.A., Wheatley, M.J. & Holloway, M.G. (1997a). Comparison of fish assemblages associated with seagrass and adjacent unvegetated habitats of Port Phillip Bay and Corner Inlet, Victoria, Australia, with emphasis on commercial species. *Estuarine, Coastal and Shelf Science*, **44**, 469–588.

Jenkins, G.P., Black, K.P., Wheatley, M.J. & Hatton, D.N. (1997b). Temporal and spatial variability in recruitment of a temperate, seagrass-associated fish is largely determined by physical processes in the pre- and post-settlement phases. *Marine Ecology Progress Series*, **148**, 23–35.

Jernakoff, P. & Nielsen, J. (1997). The relative importance of amphipod and gastropod grazers in *Posidonia sinuosa* meadows. *Aquatic Botany*, **56**, 183–202.

Jernakoff, P., Brearly, A. & Nielsen, J. (1996). Factors affecting grazer-epiphyte interactions in temperate seagrass meadows. *Oceanography and Marine Biology: an Annual Review*, **34**, 109–62.

Jordan, F., Bartolini, M., Nelson, C., Patterson, P.E. & Soulen, H.L. (1996). Risk of predation affects habitat selection by the pinfish *Lagodon rhomboides* (Linnaeus). *Journal of Experimental Marine Biology and Ecology*, **208**, 45–56.

Judge, M.L., Coen, L.D. & Heck, K.L., Jr. (1993). Does *Mercenaria mercenaria* encounter elevated food levels in seagrass beds? Results from a novel technique to collect suspended food resources. *Marine Ecology Progress Series*, **141**, 141–50.

Kenyon, R.A., Loneragan, N.R. & Hughes, J.M. (1995). Habitat type and light affect sheltering behaviour of juvenile tiger prawns (*Penaeus esculentus* Haswell) and success rates of their fish predators. *Journal of Experimental Marine Biology and Ecology*, **192**, 87–105.

Kenyon, R.A., Loneragan, N.R., Hughes, J.M. & Staples, D.J. (1997). Habitat type influences the microhabitat preference of juvenile tiger prawns (*Penaeus esculentus* Haswell and *Penaeus semisulcatus* De Haan). *Estuarine, Coastal and Shelf Science*, **45**, 393–403.

Kikuchi, T. (1966). An ecological study on animal communities of the *Zostera marina* belt in Tomioka Bay, Amakusa, Kyushu. *Publ. Amakusa Mar. Biol. Lab.*, **1**, 127–48.

Klumpp, D.W., Howard, R.K. & Pollard, D.A. (1989). Trophodynamics and nutritional ecology of seagrass communities. In *Biology of Seagrasses*, ed. A.W.D Larkum, A.J. McComb & S.A. Shepherd, pp. 394–457. Amsterdam: Elsevier.

Laegdsgaard, P. & Johnson, C.R. (1995). Mangrove habitats as nurseries: unique assemblages of juvenile fish in subtropical mangroves in eastern Australia. *Marine Ecology Progress Series*, **126**, 67–81.

Lanyon, J.M., Limpus, C.J. & Marsh, H. (1989). Dugongs and turtles: grazers in the seagrass system. In *Biology of Seagrasses*, ed. A.W.D. Larkum, A.J. McComb & S.A. Shepherd, pp. 610–34. Amsterdam: Elsevier.

Liu, H. & Loneragan, N.R. (1997). Size and time of day affect the response of postlarvae and early juvenile grooved tiger prawns *Penaeus semisulcatus* De Haan (Decapoda: Penaeidae) to natural and artificial seagrass in the laboratory. *Journal of Experimental Marine Biology and Ecology*, **211**, 262–77.

Loneragan, N.R., Kenyon, R.A., Haywood, M.D.E. & Staples, D.J. (1994). Population dynamics of juvenile tiger prawns (*Penaeus esculentus* and *P. semisulcatus*) in seagrass habitats of the western Gulf of Carpentaria, Australia. *Marine Biology*, **119**, 133–43.

Loneragan, N.R., Kenyon, R.A., Staples, D.J., Poiner, I.R. & Conacher, C.A. (1998). The influence of seagrass type on the distribution and abundance of postlarval and juvenile tiger prawns (*Penaeus esculentus* and *P. semisulcatus*) in the western Gulf of Carpentaria, Australia. *Journal of Experimental Marine Biology and Ecology*, **228**, 175–95.

Marsh, H., Channells, P.W., Heinsohn, G.E. & Morrissey, J. (1982). Analysis of stomach contents of dugongs from Queensland. *Australian Wildlife Research*, **9**, 55–67.

McMillan, R.O., Armstrong, D.A. & Dinnel, P.A. (1995). Comparison of intertidal habitat use and growth rates of two northern Puget Sound cohorts of 0+ age Dungeness crab, *Cancer magister*. *Estuaries*, **18**, 390–8.

Moksnes, P.-O., Lipcius, R., Pihl, L. & van Montfrans, J. (1997). Cannibal-prey dynamics in young juveniles and postlarvae of the blue crab. *Journal of Experimental Marine Biology and Ecology*, **215**, 157–87.

Moreteau, J.C. & Vicente, N. (1982). Evolution d'une population de *Pinna nobilis* L. (Mollusca, Bivalvia). *Malacologia*, **22**, 341–5.

Mortimer, J.A. (1981). The feeding ecology of the West Caribbean green turtle, *Chelonia mydas*, in Nicaragua. *Biotropica*, **13**, 49–58.

Mortimer, J.A. (1995). Feeding ecology of sea turtles. In *Biology and Conservation of Sea Turtles*, ed. K.A. Bjorndal, pp. 103–9. Washington DC: Smithsonian Institution Press.

Motta, P.J., Clifton, K.B., Hernandez, P., Eggold, B.T., Giordano, S.D. &

Wilcox, R. (1995). Feeding relationships among nine species of seagrass fishes of Tampa Bay, Florida. *Bulletin of Marine Science*, **56**, 185–200.

Neckles, H.A., Wetzel, R.L. & Orth, R.J. (1993). Relative effects of nutrient enrichment and grazing on epiphyte-macrophyte (*Zostera marina* L.) dynamics. *Oecologia*, **93**, 285–95.

Ogden, J.C., Robinson, L., Whitlock, K., Dagenhart, H. & Cebula, R. (1983). Diel foraging patterns in juvenile green turtles (*Chelonia mydas* L.) in St. Croix, United States Virgin Islands. *Journal of Experimental Marine Biology and Ecology*, **66**, 199–205.

Olmi, E.J. III & Lipcius, R.N. (1991). Predation on postlarvae of the blue crab *Callinectes sapidus* Rathbun by sand shrimp *Crangon septemspinosa* Say and grass shrimp *Palaemonetes pugio* Holthuis. *Journal of Experimental Marine Biology and Ecology*, **151**, 169–83.

Orth, R.J., Heck, K.L., Jr. & van Montfrans, J. (1984). Faunal communities in seagrass beds: a review of the influence of plant structure and prey characteristics on predator-prey relationships. *Estuaries*, **7**, 339–50.

Orth, R.J., van Montfrans, J., Lipcius, R.N. & Metcalf, K.S. (1996). Utilization of seagrass habitat by the Blue Crab, *Callinectes sapidus* Rathbun, in Chesapeake Bay: a review. In *Seagrass Biology: Proceedings of an international workshop*, ed. J. Kuo, R.C. Phillips, D.I. Walker & H. Kirkman, pp. 213–24. Perth: University of Western Australia.

Philippart, C.J.M. (1995). Effect of periphyton grazing by *Hydrobia ulvae* on the growth of *Zostera noltii* on a tidal flat in the Dutch Wadden Sea. *Marine Biology*, **122**, 431–37.

Pinto, L. & Punchihewa, N.N. (1996). Utilisation of mangroves and seagrasses by fishes in the Negombo Estuary, Sri Lanka. *Marine Biology*, **126**, 333–45.

Pohle, D.G., Bricelj, V.M. & García-Esquivel, Z. (1991). The eelgrass canopy: an above-bottom refuge from benthic predators for juvenile bay scallops *Argopecten irradians*. *Marine Ecology Progress Series*, **74**, 47–59.

Pollard, D.A. (1984). A review of ecological studies on seagrass-fish communities, with particular reference to recent studies in Australia. *Aquatic Botany*, **18**, 3–42.

Powell, J.A. (1978). Evidence of carnivory in manatees (*Trichechus manatus*). *Journal of Mammology*, **59**, 44.

Preen, A. R. (1993). Interactions between dugongs and seagrasses in a subtropical environment. Ph.D. Thesis, James Cook University of North Queenland, Australia.

Preen, A. (1995a). Diet of dugongs: are they omnivores? *Journal of Mammology*, **76**, 163–71.

Preen, A. (1995b). Impacts of dugong foraging on seagrass habitats: observational and experimental evidence for cultivation grazing. *Marine Ecology Progress Series*, **124**, 201–13.

Preen, A. & Marsh, H. (1995). Response of dugongs to large-scale loss of seagrass from Hervey Bay, Queensland, Australia. *Wildlife Research*, **22**, 507–19.

Preen, A.R., Lee Long, W.J. & Coles, R.G. (1995). Flood and cyclone related loss, and partial recovery, of more than 1000 km^2 of seagrass in Hervey Bay, Queensland, Australia. *Aquatic Botany*, **52**, 3–17.

Reusch, T.B.H. (1998). Differing effects of eelgrass *Zostera marina* on recruitment and growth of associated blue mussels *Mytilus edulis*. *Marine Ecology Progress Series*, **167**, 149–53.

Reusch, T.B.H. & Chapman, A.R.O. (1995). Storm effects on eelgrass (*Zostera marina*) and blue mussel (*Mytilus edulis* L.) beds. *Journal of Experimental Marine Biology and Ecology*, **192**, 257–71.

Reusch, T.B.H. & Williams, S.L. (1998). Variable responses of native eelgrass *Zostera marina* to a non-indigenous bivalve *Musculista senhousia*. *Oecologia*, **113**, 428–41.

Reusch, T.B.H., Chapman, A.R.O. & Gröger, J.P. (1994). Blue mussel *Mytilus edulis* do not interfere with eelgrass *Zostera marina* but fertilize shoot growth through biodeposition. *Marine Ecology Progress Series*, **108**, 265–82.

Richardson, C.A., Kennedy, H.A., Duarte, C.M., Kennedy, D.P. & Proud, S.V. (1999). Population density and growth of the fan mussel, *Pinna nobilis* from S.E. Spanish Mediterranean seagrass, *Posidonia oceanica*, meadows. *Marine Biology*, **133**, 205–212.

Robblee, M.B. & Zieman, J.C. (1984). Diel variation in the fish fauna of a tropical seagrass feeding ground. *Bulletin of Marine Science*, **34**, 335–45.

Rooker, J.R., Holt, S.A., Soto, M.A. & Holt, G.J. (1998). Postsettlement patterns of habitat use by Sciaenid fishes in subtropical seagrass meadows. *Estuaries*, **21**, 318–27.

Ross, S.W. & Moser, M.L. (1995). Life history of juvenile gag, *Mycteroperca microlepis*, in North Carolina estuaries. *Bulletin of Marine Science*, **56**, 222–37.

Rozas, L.P. & Minello, T.J. (1998). Nekton use of salt marsh, seagrass, and nonvegetated habitats in a South Texas (USA) estuary. *Bulletin of Marine Science*, **63**, 481–501.

Sedberry, G.R. & Carter, J. (1993). The fish community of a shallow tropical lagoon in Belize, Central America. *Estuaries*, **16**, 198–215.

Sheridan, P.F. (1992). Comparative habitat utilization by estuarine macrofauna within the mangrove ecosystem of Rookery Bay, Florida. *Bulletin of Marine Science*, **50**, 21–39.

Sheridan, P. (1997). Benthos of adjacent mangrove, seagrass and non-vegetated habitats in Rookery Bay, Florida, U.S.A. *Estuarine, Coastal and Shelf Science*, **44**, 455–69.

Sogard, S.M. (1992). Variability in growth rates of juvenile fishes in different estuarine habitats. *Marine Ecology Progress Series*, **85**, 35–53.

Sogard, S.M. & Able, K.W. (1991). A comparison of eelgrass, sea lettuce, macroalgae, and marsh creeks as habitats for epibenthic fishes and decapods. *Estuarine, Coastal and Shelf Science*, **33**, 501–19.

Sogard, S.M., Powell, G.V.N. & Holmquist, J.G. (1987). Epibenthic fish communities on Florida Bay banks: relations with physical parameters and seagrass cover. *Marine Ecology Progress Series*, **40**, 25–39.

Somaschini, A., Gravina, M.F. & Ardizzone, G.D. (1994). Polychaete depth distribution in a *Posidonia oceanica* bed (rhizome and matte strata) and neighbouring soft and hard bottoms. *P.S.Z.N. I: Marine Ecology*, **15**, 133–51.

Stapel, J. (1997). Nutrient dynamics in Indonesian seagrass beds: factors determining conservation and loss of nitrogen and phosphorus. Ph.D. Thesis, Catholic University Nijmegen, Netherlands.

Stapel, J. & Hemminga, M.A. (1997). Nutrient resorption from seagrass leaves. *Marine Biology*, **128**, 197–206.

Suchanek, T.H. (1983). Control of seagrass communities and sediment distribution by *Callianassa* (Crustacea, Thalassinidae) bioturbation. *Journal of Marine Research*, **41**, 281–98.

Tolan, J.M., Holt, S.A. & Onuf, C.P. (1997). Distribution and community structure of ichtyoplankton in Laguna Madre seagrass meadows: potential impact of seagrass species change. *Estuaries*, **20**, 450–64.

Valentine, J.F., Heck, K.L., Jr., Harper, P. & Beck, M. (1994). Effects of bioturbation in controlling turtlegrass (*Thalassia testudinum* Banks ex König) abundance: evidence from field enclosures and observations in the Northern Gulf of Mexico. *Journal of Experimental Marine Biology and Ecology*, **178**, 181–92.

Valentine, J.F., Heck, K.L., Jr., Busby, J. & Webb, D. (1997). Experimental evidence that herbivory increases shoot density and productivity in a subtropical turtlegrass (*Thalassia testudinum*) meadow. *Oecologia*, **112**, 193–200.

Vance, D.J., Haywood, M.D.E., Heales, D.S. & Staples, D.J. (1996). Seasonal and annual variation in abundance of postlarval and juvenile grooved tiger prawns *Penaeus semisulcatus* and environmental variation in the Embley River, Australia: a six year study. *Marine Ecology Progress Series*, **135**, 43–55.

Van Montfrans, J., Wetzel, R.L. & Orth, R.J. (1984). Epiphyte-grazer relationships in seagrass meadows: consequences for seagrass growth and production. *Estuaries*, **7**, 289–309.

Walsh, C.J. & Mitchell, B.D. (1998). Factors associated with variations in abundance of epifaunal caridean shrimps between and within estuarine seagrass meadows. *Marine and Freshwater Research*, **49**, 769–77.

Weinstein, M.P. & Heck, K.L., Jr. (1979). Ichtyofauna of seagrass meadows along the Caribbean coast of Panama and in the Gulf of Mexico: composition, structure and community ecology. *Marine Biology*, **50**, 97–107.

Whiting, S.D. & Miller, J.D. (1998). Short term foraging ranges of adult green turtles (*Chelonia mydas*). *Journal of Herpetology*, **32**, 330–7.

Williams, S.L. (1988). *Thalassia testudinum* productivity and grazing by green turtles in a highly disturbed seagrass bed. *Marine Biology*, **98**, 447–55.

Williams, S.L. & Ruckelshaus, M.H. (1993). Effects of nitrogen availability and herbivory on eelgrass (*Zostera marina*) and epiphytes. *Ecology*, **74**, 904–18.

Woods, C.M.C. & Schiel, D.R. (1997). Use of seagrass *Zostera novazelandica* (Setchell, 1993) as habitat and food by the crab *Macrophtalmus hirtipes* (Heller, 1862) (Brachyura: Ocypodidae) on rocky intertidal platforms in southern New Zealand. *Journal of Experimental Marine Biology and Ecology*, **214**, 49–65.

Worthington, D.G., McNeill, S.E., Ferrell, D.J. & Bell, J.D. (1995). Large scale variation in abundance of five common species of decapod sampled from seagrass in New South Wales. *Australian Journal of Ecology*, **20**, 515–25.

Zieman, J.C., Iverson, R.L. & Ogden, J.C. (1984). Herbivory effects on *Thalassia testudinum* leaf growth and nitrogen content. *Marine Ecology Progress Series*, **15**, 151–8.

Zimmerman, R.C., Kohrs, D.G. & Alberte, R.S. (1996). Top-down impact through a bottom-up mechanism: the effect of limpet grazing on growth, productivity and carbon allocation of *Zostera marina* L. (eelgrass). *Oecologia*, **107**, 560–7.

7

Seagrasses in the human environment

7.1 Introduction

The rapidly expanding scientific knowledge on seagrasses has led to a growing awareness that seagrasses are a valuable coastal resource. Where seagrasses abound, humans benefit directly and indirectly from the presence of this marine vegetation. At the same time, it has also become evident that seagrasses are a *vulnerable* resource, easily lost in coastal areas facing environmental changes. Declines of seagrasses are reported world-wide, and in many cases anthropogenic factors are suspected to be responsible for these declines. In this chapter the relations between seagrasses and humans are addressed. Particular attention will be given to the various stresses on seagrasses resulting from human activities. Knowledge of the processes that lead to seagrass decline is obviously the key to remedial measures targeting the re-establishment or protection of seagrass systems.

7.2 The value of seagrasses to humans

The value of seagrasses, as perceived by humans, changes in time and place. In the past, seagrasses have been valued because the plants yielded material for various practical purposes. This direct use of seagrasses has a long history that continues, although on a very modest scale, until today. Seventeenth and eighteenth century Spanish colonial documents indicate that the seeds of *Zostera marina* were a major food resource of the Seri Indians living along the Gulf of California. The Seri harvested the carbohydrate-rich seeds to obtain flour that was used in different dishes (Felger *et al.*, 1980). In the north-west Pacific, roots and leaf bases of eelgrass were eaten (Turner & Bell, 1973). Human consumption of seagrass biomass is not entirely confined to the past. In South-East Asia, seeds of *Enhalus acoroides* are still a food source for coastal populations. The nutritional value of the flour derived from the *Enhalus* seeds is comparable to that of wheat and rice flour in terms of carbohydrate and protein content and in energetic value, and surpasses these types of flour in calcium, iron and phosphorus content (Montaño *et al.*, 1999).

Archaeological evidence suggests that in medieval Europe *Zostera marina* was burned to obtain salt (B. van Geel *et al.*, unpublished data), a practice that continued until the nineteenth century, at least in Denmark (Brøndegaard, 1987). Old sources also mention the widespread use of *Zostera* as filling material for pillows and matresses, as roofing material, and as fodder in the north-west. European countries. In the Netherlands, an important application of eelgrass was its use in dike building: layers of seagrass material protected the dikes against wave action. Large-scale harvesting of eelgrass continued until the first decades of the twentieth century. In this last period, the United States even imported eelgrass for use in insulation for sound and temperature control (McRoy & Helfferich, 1980). Presently, eelgrass is still extensively harvested in the Aveiro lagoon (Portugal) for use as soil fertilizer.

In southern Europe, beach-cast *Posidonia oceanica* leaves have been used for a number of purposes in past centuries, including use as shock-absorbing material for the transport of glassware, to keep fish catches moist during transportation, as filling material for matresses, and as compost and fertilizer in the fields (Font-Quer, 1980). In Australia small-scale harvesting of beach-cast leaves of *Posidonia* spp., for the production of soil improvers, mulch and compost, continued into the 1990s in South Australia (Kirkman & Kendrick, 1997). *Posidonia* leaves were also used as house insulation. Beach-cast leaves of *Zostera* spp. and *Heterozostera tasmanica* were harvested for a similar purpose, and also for soil improver in south-east Australia, but commercial harvesting is nowadays prohibited (Kirkman & Kendrick, 1997).

Current appreciation of seagrasses concerns not so much the direct use value, but the services that seagrass vegetation provide to overall functioning of coastal zone systems:

(1) Seagrass meadows enhance the biodiversity and habitat diversity of coastal waters. Seagrasses are remarkable in being the only truly marine angiosperms. Their presence therefore implies a direct and unique enrichment of the variety of life forms in coastal waters. Moreover, for a variety of epiphytic algae and

faunal organisms, the seagrass vegetation offers a suitable habitat (section 6.2). A few of these are exclusively dependent on seagrasses, and are lacking where seagrasses are absent. Most can live outside the seagrass meadows as well, but the possibilities for attachment, the refuge value of the meadows and the diversity of food resources that is found within them, leads to a concentration of species and an abundance of individuals not associated with bare areas.

(2) Seagrass meadows support the production of living marine resources: they are nursery and foraging areas for a number of commercially highly important fish and shellfish species (sections 6.3 and 6.4). As discussed, not every meadow has a similar value, but sometimes their direct economical importance is undisputed. The considerable dependence of tropical penaeid prawns on the seagrass habitat is an example. A reduction in the areal distribution of seagrass vegetation that harbours the juveniles of commercially important penaeids is expected to have direct consequences for the adult prawn stocks that are fished in offshore waters. In the case of the seagrass meadows (876 ha) in Cairns Harbour, Australia, it was calculated that the potential annual yield of the three major commercial prawn species from these meadows amounted to between 0.6 and 2.2 million Australian dollars per year (Watson *et al.*, 1993).

(3) Seagrasses meadows improve water quality by reducing particle loads in the water and absorbing dissolved nutrients. Seagrass meadows act as roughness elements that deflect currents and dissipate the kinetic energy of the water, thereby creating a relatively quiet environment favourable for sediment deposition and retention (section 5.6.3). Associated suspension feeders, furthermore, can greatly contribute to the particle-trapping character of seagrass meadows. The significance of these seagrass-associated suspension feeders for water quality is only beginning to emerge, but studies on suspension feeders in non-vegetated systems indicate that these can be highly effective in dampening increases in phytoplankton biomass following nutrient additions to the water (Herman & Scholten, 1990). In addition, seagrasses effectively capture dissolved nutrients by their leaves (section 4.4.3), and sequester these in their tissues. Once the nutrients are fixed in the seagrass biomass, they are returned to the water column only after a decomposition process

that is slow compared with the decomposition of phytoplankton cells. By trapping nutrients in slowly degradable tissues, the possibilities for rapid growth and turnover of phytoplankton cells is reduced. It is interesting to note that in an assessment of the economic value of the the world's ecosystems, the value of the nutrient cycling function of seagrass beds and beds of macroalgae, expressed in dollars per ha per year, ranked second among the listed ecosystem values (Costanza *et al.*, 1997).

(4) Seagrass stabilize sediments. The reduced flow regime inside seagrass canopies diminishes sediment resuspension. Seagrasses also develop a web of roots that binds the sediment, and stabilizes it. In addition, some seagrass species, notably *Posidonia oceanica*, form reef-like structures reaching close to the water surface that dissipate the wave energy before it reaches the shoreline (Molinier & Picard, 1952). The role of seagrasses in coastal protection is best documented by the major episodes of coastal erosion and shoreline regression that followed eelgrass decline (Christiansen *et al.*, 1981). This role is not restricted to the larger seagrass species, for even the tiny canopy of *Halophila* species provides protection of the sediments against resuspension (Fonseca, 1989). Even beach-cast seagrass material may stabilize shorelines, by trapping sand and inducing dune formation (Hemminga & Nieuwenhuize, 1990).

(5) Seagrasses play a significant role in global carbon and nutrient cycling, and awareness of this aspect of the value of seagrasses is recently emerging. Seagrass biomass has already been identified, together with that of long-lived macroalgae, as an important sink for carbon in the ocean (Smith, 1981). Most of the biomass produced by seagrasses ends up as refractory detritus. As a result, the quantity of seagrass carbon available to be stored in the sediments is large, representing some 0.08 Pg C year^{-1} (1 Pg = 10^{15} g) in the ocean as a whole. This represents about 12% of the total carbon storage in the ocean, despite the production of seagrasses representing only a small percentage (1%) of the total oceanic production (Duarte & Cebrián, 1996; Table 7.1). Because of the relatively low nutrient concentration in seagrass tissues, the burial of nitrogen and phosphorus associated with the burial of seagrass detritus could be expected to be less important. This is, however, not the case, for the refractory nature of seagrass detritus implies that its burial efficiency

Table 7.1. *Estimates of total net primary production (NPP) by different primary producers and the amount of this production that is consumed by herbivores, decomposed, or stored*

The stored fraction represents the detritus that is not decomposed within a year, and, hence, accumulates annually in the system. Values in parentheses are percentages of total herbivory, decomposition or storage in the ocean.

Primary producer	Area covered (10^6 km^2)	Total NPP (Pg C yr^{-1})	Herbivory (Pg C yr^{-1})	Decomposition (Pg C yr^{-1})	Storage (Pg C yr^{-1})
Oceanic phytoplankton	332	43	24.4 (88)	14.7 (77.5)	0.17 (26.5)
Coastal phytoplankton	27	4.5	1.8 (6.5)	1.8 (9.8)	0.18 (27.0)
Microphytobenthos	6.8	0.34	0.15 (0.5)	0.09 (0.4)	0.02 (3.1)
Coral reef algae	0.6	0.6	0.18 (0.6)	0.45 (2.0)	0 (0.7)
Macroalgae	6.8	2.55	0.86 (3.1)	0.95 (4.2)	0.01 (1.6)
Seagrasses	0.6	0.49	0.09 (0.3)	0.25 (1.1)	0.08 (12.0)
Marsh plants	0.4	0.44	0.14 (0.5)	0.23 (1.0)	0.07 (11.3)
Mangroves	1.1	1.1	0.10 (0.3)	0.44 (1.9)	0.11 (17.6)
Total		53	27.8 (52)	19.0 (36)	0.65 (1.2)

Source: Duarte & Cebriän (1996).

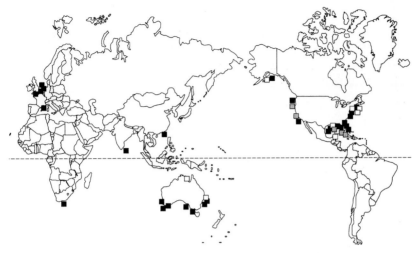

Fig. 7.1. Seagrass declines reported since the 1970s. White symbols indicate natural declines; black symbols indicate anthropogenically caused declines, and shaded symbols show declines where the cause has not established with certainty as to be either natural or anthropogenic. (Figure based on the data compiled by Short & Wyllie-Echeverria, 1996, and augmented with observations of Hine *et al.*, 1987, Lee, 1994, and Preen *et al.*, 1995.)

(i.e. the percentage of the mass stored in the sediments that is not remineralized through diagenetic processes) is likely to exceed that of most other marine detritus. In addition to burial of nutrients, seagrass meadows can also represent important entry points for nitrogen – through nitrogen fixation – at the regional scale. Bethoux & Copin-Montégut (1986), for instance, concluded that nitrogen fixation associated with *Posidonia oceanica* meadows annually accounts for an input of 57×10^{10} g N in the Mediterranean.

7.3 The occurrence of large-scale declines

Since the rapid proliferation of seagrass research in the 1970s, many reports of large-scale declines of seagrass areas have been published. Fig. 7.1 shows the locations in the world where, since the 1970s, such declines have been documented. The figure is largely based on data compiled by Short & Wyllie-Echeverria (1996), augmented with some data from other authors. To date, at more than 40 locations large-scale declines, involving at least 24 different species of seagrass, have been documented. A whole array of natural and human-induced disturbances are the cause of these declines. In a small minority of the investigated cases, natural disturbances led to population declines; in more than 70% of the cases, however, human-induced disturbances are held responsible (Short & Wyllie-Echeverria, 1996).

7.4 Natural causes of decline

Natural causes of large-scale declines comprise geological and meteorological events, and specific biological interactions. An example of the first category is provided by the decline of *Zostera marina* beds after the great earthquake of 1964 in Alaska. This earthquake caused a rise of the shoreline, and subsequent exposure of the eelgrass vegetation (Johansen, 1971, cited in Short & Wyllie-Echeverria, 1996). Anecdotal information indicates that the ash and debris rains that accompany volcanic eruptions may eradicate seagrass beds as well (Short & Wyllie-Echeverria, 1996). Hurricanes and cyclones repeatedly have been observed to cause widespread damage to submerged vegetation, although they may also pass without causing much damage (Poiner *et al.*, 1989). These windstorms coincide with huge waves and strong currents over the seafloor, which may uproot seagrasses and erode the sediment surface. Investigation of seagrass communities shortly after the passage of a cyclone in 1985 over the western Gulf of Carpentaria (Australia) showed that of the 183 km^2 of seagrasses recorded in the area, some 150 km^2 had been lost, undermined or smothered by thick layers of fine mud (Poiner *et al.*, 1989). More recently, an unusual combination of two floods and a cyclone, causing a prolonged episode of highly turbid coastal water, resulted in a 1000 km^2 loss of seagrass in north-eastern Australia (Preen *et al.*, 1995). The activities of humans may sometimes exacerbate the effects of natural disturbances. In the forementioned case, increased soil erosion following forest clearing and overgrazing by livestock probably aggravated turbidity in the flood plumes (Preen *et al.*, 1995). In the case of the declines of *Posidonia australis* and *Zostera capricorni* in Botany Bay (Australia), dredging of the bay's entrance enhanced storm effects. Dredging resulted in increased wave heights in the bay, causing excessive erosion of the beds during severe storms (Larkum & West, 1990).

Declines of seagrass beds can be part of a natural cycle of meadow development and breakdown at time scales of several years or decades, as described for some *Zostera marina* populations by Glémarec (1979) and Den Hartog (1987). Intertidal *Zostera marina* beds along the coasts of north-western Europe, for example, show considerable variation in developmental cycles of the beds. Most beds typically display an annual cycle. The vegetation disappears after the growing season: grazing by waterfowl, mortality of the plants due to low winter temperatures, and erosion of the beds due to floating ice, eradicate the beds each year. The beds develop again the next spring from seeds. This growth cycle coincides with a seasonal cycle of sedimentation in the months that vegetation is present, and sediment erosion during exposure of the surface in winter. On an annual basis, sedimentation and erosion are more or less in balance, maintaining the beds at a constant elevation in the tidal range. A quite different situation, however, is found at some places. In areas where frost is exceptional, and grazing pressure is low, such as in the area near Roscoff in the western part of the English Channel, the meadows may persist throughout the year. This allows continuous accretion of the beds over periods of years, ultimately raising them to levels that are less favourable to the plants. Eventually, after years of vigorous growth, decline of these elevated beds sets in. A new growth phase is expected again only after sufficient lowering of the sediment surface (Den Hartog, 1987).

The 'wasting disease' of the early 1930s, which led to a dramatic decline of *Zostera marina* in coastal waters of Europe and North America, undoubtedly is the most notorious example of a biological interaction that results in seagrass loss. The organism responsible has been identified as *Labyrinthula zosterae* (Muehlstein *et al.*, 1991). This is a representative of the Labyrinthulaceae, or marine slime moulds. The symptoms of the wasting disease in eelgrass are quite characteristic. On the leaves small dark brown or black lesions appear, which spread longitudinally over the leaves, eventually covering them entirely after a few weeks (Short *et al.*, 1988; Vergeer & Den Hartog, 1994). These symptoms may be restricted to mature leaves, but in severely affected plants they are even apparent in the youngest leaves. An outbreak of the disease on the scale of the 1930s has not been repeated since, but scattered outbreaks of the disease continue to affect *Z. marina* populations on local scales (Short *et al.*, 1986, 1988). *Labyrinthula zosterae* can infect *Zostera noltii* as well, causing a similar type of leaf damage as in *Z. marina* (Vergeer & Den Hartog, 1991), but large-scale deterioration of *Z. noltii* meadows due to wasting disease has not been recorded so far. Symptoms of wasting disease, caused by an unidentified *Labyrinthula* species have also been observed in *Thalassia testudinum* (Durako & Kuss, 1994), and this slime mould may have contributed to the seagrass decline that started in the late 1980s in Florida Bay (Porter & Muehlstein, 1989). It is now clear that species belonging to the genus *Labyrinthula* occur widely on different seagrasses from all parts of the world, apparently without causing harm to the plants (Vergeer & Den Hartog, 1994). Probably, the normal association between plants and (some) slime moulds may only get a pathogenous character under specific conditions, e.g. when environmental stresses (natural or human-induced) have made the plants more vulnerable (Den Hartog, 1987; Short *et al.*, 1988; Vergeer & Den Hartog, 1994). Low irradiance conditions coinciding with increased water turbidity, for instance, probably reduced the vitality of eelgrass in the Dutch Wadden Sea during the growing seasons of 1931 and 1932, making the plants susceptible to the wasting disease that annihilated the region's subtidal meadows (Giesen *et al.*, 1990).

Areal declines have also been reported as a result of animal activity. A population explosion of herbivorous sea urchins in *Posidonia australis* meadows in Botany Bay (Australia) resulted in the complete removal of the seagrass canopy over areas of tens of hectares. The plants did not have the resilience to recover from this massive grazing event; several years later, litter from roots and rhizomes were the only remains of the meadows (Larkum & West, 1990). In Monterey Bay, California, high densities of the small limpet *Tectura depicta*, which consumes the chloroplast-containing epidermis of *Zostera marina*, reduced a 30 ha meadow to less than 50% of its original size (Zimmerman *et al.*, 1996). Not only direct grazing activity of animals may lead to meadow deterioration. In the Dutch Wadden Sea, an increase of the density of lugworms (*Arenicola marina*) is believed to be responsible for local declines of *Zostera noltii*. The bioturbation activity of these animals results in coverage of the shoots of this small seagrass by layers of sediment, a process that, at high lugworm densities, has a devastating effect (Philippart, 1994).

7.5 Anthropogenic causes of declines

7.5.1 *Eutrophication*

Eutrophication is a serious problem in the world's coastal zones (Nixon, 1995; Vidal *et al.*, 1999). Worldwide population growth, together with the massive use of fertilizers in agriculture, have led to an exponential increase in nutrient inputs to the coastal zone (Nixon, 1995), with areal loads sometimes being comparable to those of heavily fertilized agricultural fields (Borum, 1996). Symptoms of eutrophication have been

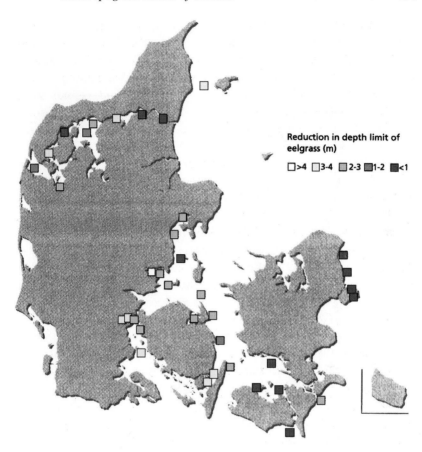

Fig. 7.2. The reduction in depth limit of *Zostera marina* along the Danish coast since the turn of the century. (Christensen *et al.*, 1995.)

documented in coastal areas all around the world, and these symptoms are also apparent in waters which appear still to be rather oligotrophic, such as the coastal waters of the Mediterranean, where transparency measured as Secchi depth has been declining on average 0.1 m y^{-1} since 1975 (Marbà & Duarte, 1997). Indeed, the deterioration of water transparency associated with eutrophication is no longer a local problem: it affects the coastal waters of entire nations, causing widespread seagrass decline (Fig. 7.2). The wealth of reports on seagrass decline following eutrophication renders the negative effect of marine eutrophication on seagrass stands an indisputable fact, and indicates that it most likely is the main cause of seagrass decline worldwide. For

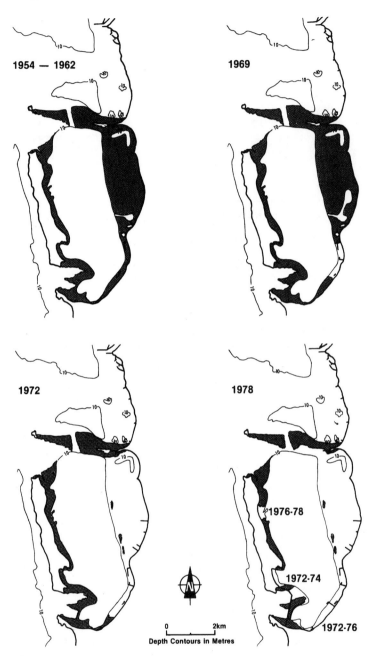

Fig. 7.3. The time course of seagrass loss in Cockburn Sound (Western Australia) during a period of industrial development. Seagrass areas are shaded. (Cambridge & McComb, 1984.)

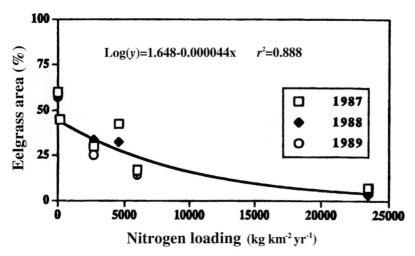

Fig. 7.4. The relationship between eelgrass area and nitrogen loading to Waquoit Bay (Massachusetts, USA). (Short & Burdick, 1996.)

instance, Cambridge & McComb (1984) documented the loss of 80% of the seagrass area in Cockburn Sound (Western Australia) between 1954 and 1978, parallel to a period of industrial development (Fig. 7.3). Short & Burdick (1996) provided evidence of strong negative correlations between the area occupied by *Zostera marina* in Waquoit Bay (Massachusetts, USA) and nitrogen loading, which was, in turn, closely associated with housing development (Fig. 7.4). Strong, negative correlations between the depth limit of eelgrass (*Zostera marina*) and total nitrogen concentrations in the water column (Fig. 7.5) have further provided evidence for the negative effects of increased nutrient supply on seagrass stands (Borum, 1996).

Marine primary producers are often nutrient limited, and in particular nitrogen limitation is common (Downing et al., 1999). This leads to a general stimulation of their growth following nutrient inputs. Seagrasses are also often nutrient limited (see section 4.4.4), and they therefore could be expected to benefit from an enhanced nutrient supply, except perhaps supplies that would result in very high water column concentrations, as there is some evidence that strongly elevated nitrate and ammonium concentrations can be directly toxic to seagrasses (Burkholder et al., 1992, 1994; Van Katwijk et al., 1997). Seagrass decline following eutrophication therefore generally cannot be explained on the basis of direct effects of the elevated nutrient concentrations themselves, but on

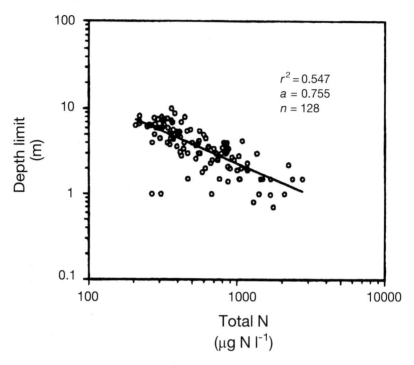

Fig. 7.5. The relationship between the depth limit of *Zostera marina* in Danish coastal waters and the total nitrogen concentration. (Borum, 1996.)

indirect effects associated with the far greater response of other marine primary producers.

Seagrasses are able to achieve high primary production in oligotrophic environments, such as many tropical waters, because, in contrast to most other marine primary producers, they can exploit sediment nutrient pools. Seagrasses also have internal nutrient concentrations well below those of other marine primary producers (Duarte, 1992). In combination with their comparatively slower growth rates (Duarte, 1995; Nielsen *et al.*, 1996), this results in nutrient requirements per weight unit that are about 50–100-fold lower and 1.5–8-fold lower than those of phytoplankton and macroalgae, respectively (Duarte, 1995). Seagrasses are thus able to maintain primary production well above that of phytoplankton and macroalgae at similar nutrient inputs. Moreover, seagrasses typically experience much lower grazing losses than macroalgae and phytoplankton do (Cebrián & Duarte, 1994; Duarte, 1995), resulting in a low turnover of seagrass biomass compared with that of other marine primary producers, and leading to the much higher standing biomass of seagrass meadows. Because of this extensive biomass development, seagrasses also sequester considerable amounts of nutrients, preventing their use by other primary producers. For various reasons, seagrasses are therefore able to outcompete other marine primary producers in nutrient-poor environments.

As nutrient inputs increase and nutrient limitation is alleviated, other resources, particularly light, become limiting and the balance between marine primary producers is altered, because seagrasses are relatively poor competitors for light. Being benthic plants, seagrasses are in a less favourable position to harvest light than phytoplankton in the water column. In addition, light absorption by photosynthetic pigments is comparatively inefficient in seagrasses, compared with microalgae and most macroalgae (Agustí *et al.*, 1994). Light absorption per unit seagrass leaf weight is much less than that for microalgae and most macroalgae (Agustí *et al.*, 1994), so that seagrasses have higher light compensation points than their competitors (Enríquez *et al.*, 1996). The difference is even greater when the compensation irradiance for growth is considered, for seagrasses have to support the respiratory demands of their non-photosynthetic tissues (rhizomes and roots) as well, whereas microalgae and most macroalgae do not have similar sink organs. The high light requirements of seagrasses are best represented by the low photosynthetic capacity and efficiency relative to those of phytoplankton. These features contribute to the comparatively low growth rates of seagrasses (Fig. 7.6; Duarte, 1995). As nutrient inputs increase, microalgae and macroalgae proliferate; phytoplankton biomass in the water column increases; opportunistic macroalgae may overgrow them (Fig. 7.7), and carpets of epiphytes may develop on their leaves, reducing the light available for seagrass photosynthesis (Fig. 7.8). This may result in a negative carbon balance. The capacity to store carbohydrate reserves in the rhizomes allows seagrasses to withstand transient periods of shading, such as those associated with sporadic algal blooms, but these reserves cannot prevent mortality when light reduction is imposed over longer time spans (Gordon *et al.*, 1994; Onuf, 1996). Light reduction, hence, is the single most important factor responsible for seagrass decline in eutrophied waters.

The predicted increase in the biomass of epiphytes and opportunistic macroalgae with increased nutrient inputs has been confirmed by observations (Borum, 1985; Silberstein *et al.*, 1986; Tomasko & Lapointe, 1991) and experiments (Harlin & Thorne-Miller, 1981; Tomasko &

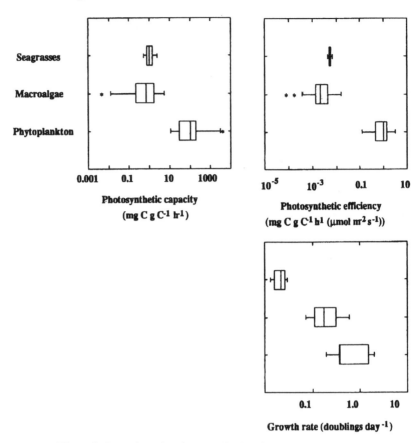

Fig. 7.6. Box plots showing the distribution of photosynthetic capacity and efficiency, and the specific growth rate for seagrasses, macroalgae, and phytoplankton. Boxes encompass the 25% and 75% quartiles of all the data for each plant type, the central line respresents the median, bars extend to the 95% confidence limits, and asterisks represent observations extending beyond the 95% confidence limits. (Duarte, 1995.)

Lapointe, 1991; McGlathery, 1995; Short *et al.*, 1995; Wear *et al.*, 1999). The epiphytic load on *Zostera marina* increased by two orders of magnitude along a eutrophication gradient in a Danish estuary (Borum, 1985). However, Coleman & Burkholder (1995) report an absence of major changes in total epiphytic biomass and production with experimental nutrient additions, but they detected a significant shift in the community structure of the epiphytic algae. Carpets of macroalgae have been reported to suffocate *Zostera marina* beds (Den Hartog, 1994), and

Fig. 7.7. An Australian seagrass bed overgrown by opportunistic algae. (Photograph by D. Walker.)

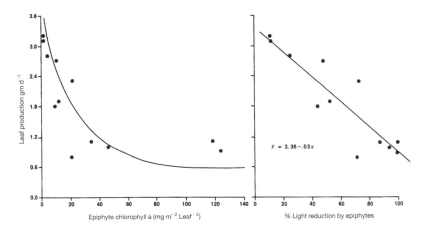

Fig. 7.8. The relationship between the epiphytic load (as the chlorophyll a density), light reduction, and the production of seagrass (*Posidonia australis*) leaves. (Silberstein *et al.*, 1986.)

may even uproot seagrasses due to the lifting power of gases trapped under the macroalgal layer covering the seagrasses. The resilience of seagrass to eutrophication may be enhanced by the presence of grazers if these are present in sufficient abundance to remove a substantial fraction of the phytoplankton production and that of the epiphytic algae (Orth &

van Montfrans, 1984; Borum, 1985; Sand-Jensen & Borum, 1991; Neckles *et al.*, 1993; Williams & Ruckelshaus, 1993). The control by grazers, however, is insufficient where reduced water exchange allows the rapid accumulation of nutrients, leading to growth rates of epiphytes that are too high to allow control by grazers. The total primary production of coastal ecosystems that harbour seagrass vegetation does not necessarily increase under eutrophied conditions, as the replacement of seagrasses by phytoplankton implies the loss of a category of autotrophs that can maintain highly productive communities at very low nutrient loadings (Borum & Sand-Jensen, 1996).

Shading by microalgae and macroalgae is not the only negative effect leading to seagrass loss, for eutrophication is also associated with increased loading of the sediments with organic matter. This raises the respiration of the benthic microbial community. As the diffusion of oxygen from the overlying water into the sediment is restricted by the low diffusivity of gases in water, the supply of oxygen as terminal electron acceptor will be thoroughly insufficient, leading to the use of alternative electron acceptors, e.g. sulphate (see section 5.4), and the development of highly reducing conditions in the sediment. Such conditions may be harmful to seagrasses, as experimental observations on some Philippine seagrass species indicate (Terrados *et al.*, 1999). The accumulation of organic matter in and on the sediments may also lead to anoxic conditions in the near-bottom water layer, particularly where strong pycnoclines exist that limit the mixing of bottom and surface waters. The development of such conditions is far more likely in summer, when high temperatures promote high microbial respiration rates and strengthen thermoclines. Consequently, losses of temperate seagrasses related to anoxic conditions are more likely in summer. High summer temperatures, furthermore, will enhance the respiration rates of the seagrasses and thus aggravate the negative effect of eutrophication-mediated light reduction on the carbon balance of the plants.

Once eutrophication has exceeded seagrass resilience, seagrass loss triggers processes that accelerate further declines (Duarte, 1995). The exposure of the sediment coinciding with the loss of seagrass vegetation tends to enhance sediment resuspension, leading to a further deterioration of light conditions for the remaining seagrass plants (Olesen, 1996). The loss of vegetation also reduces the oxygenation of the sediments and the overlying waters, further contributing to decline. The increased nutrient content and high epiphyte cover of seagrasses exposed to high nutrient loads renders them more palatable to herbivores, which have been experimentally documented to graze more intensively on sea--grasses under these conditions (McGlathery, 1995). This provides an additional mechanism for seagrass loss during the initial phases of eutrophication, although its importance may decrease subsequently due to death or migration of grazers following low oxygen conditions. As a result, seagrass decline following eutrophication can be an abrupt process, whereas recovery is certainly a slow process, as explained in Chapter 3.

Eutrophication thus may involve several environmental changes, notably a lower light availability, but also development of reduced conditions in the sediment, anoxia of bottom waters, and an increase in nutrient concentrations to potentially toxic levels, each of them having a negative effect on seagrass functioning. In addition, the response of the grazer community is important for the impact of eutrophication on seagrass vegetation. The elucidation of the relative importance of these different factors and their interactions requires an experimental approach. Burkholder *et al.* (1992) used mesocosms to test the response of *Zostera marina* to different nitrate inputs. These authors found that a decline occurred in all seasons with increasing nutrient inputs, but it was highest in periods of high temperature. Mesocosm experiments have also indicated that seagrass species differ in their susceptibility to nutrient enrichment, with fast-growing species, such as *Halodule wrightii*, being more resistant to high nutrient loading (Burkholder *et al.*, 1994). Experimental evidence suggests that negative effects are more prominent when nutrient enrichments include high nitrogen loadings, which seem to stimulate phytoplankton growth more than phosphorus additions alone (Taylor *et al.*, 1995; Lin *et al.*, 1996). Mesocosm experiments have also indicated that increased epiphyte biomass is not a consistent result of eutrophication, particularly when high phytoplankton blooms are induced (Lin *et al.*, 1996).

The link between water quality and seagrass performance is so tight that changes in seagrass cover, particularly along their depth limit, can be used as an early warning symptom of deteriorating water quality (Dennison *et al.*, 1993). Indeed, negative effects in seagrass performance correlated with high nutrient loadings have been observed before monitoring programmes were able to detect any significant deterioration in water quality (Tomasko *et al.*, 1996). Seagrasses provide, therefore, natural 'light meters' that integrate water quality parameters seasonally (Dennison *et al.*, 1993), so that they may act as robust indicator organisms in monitoring programmes.

Legislation to reduce nutrient loadings to coastal waters is progressively being implemented throughout the world, and significant reductions in nutrient inputs have already been achieved in the coastal waters of some countries. Moreover, seagrasses are increasingly protected by law in many countries. These actions should be conducive to a recovery of seagrass populations, but at present it is impossible to predict how long it will take before recovery will be complete. While seagrass loss after disturbance may be fast, recovery is likely to be slow, and simulation models based on patch formation rates and horizontal progression rates predict recovery times exceeding one century for slow growing species (Duarte, 1995). Hence, many seagrass meadows already lost will not regain their pre-disturbance distribution for several generations.

7.5.2 *Siltation*

The increasing human population is leading to changes in land use worldwide. This includes deforestation in tropical areas to produce agricultural land – often through the destructive slash and burn practice – and the clearing of land in previously unpopulated areas to provide new living space. These changes are leading, together with aggressive agricultural techniques, to high soil erosion rates and an increase in sediment transport by rivers. These effects are particularly acute in South-East Asia, where steep slopes and a high population and economic growth are leading to unparalleled deforestation rates (Panayotou, 1993). As a result, sediment delivery per unit catchment area by South-East Asian rivers is the largest in the world (Milliman & Meade, 1993), leading to widespread siltation, which threatens the coastal ecosystems in the region (Fortes, 1988; Gómez, 1988). These ecosystems comprise, as indicated earlier (Chapter 1) some of the most diverse and productive seagrass meadows in the world. Siltation presently is the main threat to these meadows (Fortes, 1988), and because of continued changes in land use, these siltation problems are even likely to increase in the future. Siltation is not confined to South-East Asia, but has also been identified as an important disturbance to seagrasses elsewhere (Kirkman & Walker, 1989; Shepherd *et al.*, 1989; Talbot *et al.*, 1990). The increased silt export to the ocean results in a deterioration of the underwater light climate for seagrasses and is also associated with increased nutrient loading (Malmer & Grip, 1994; Mitchell *et al.*, 1997). Additional negative effects on seagrasses are (partial) burial of the plants (Duarte *et al.*, 1997), and the unfavourable change of sediment conditions (e.g. redox potential)

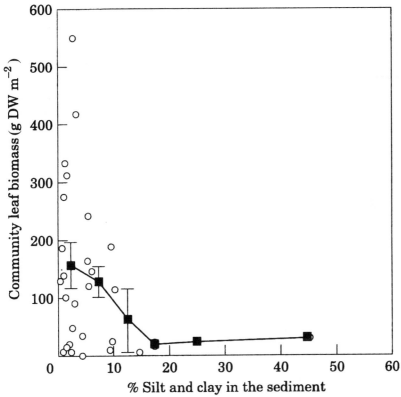

Fig. 7.9. The decline in leaf biomass of South-East Asian seagrass species with increasing silt and clay content in the sediments. Squares represent means ± SE, open symbols are individual data points. (Terrados *et al.*, 1998.)

(Terrados *et al.*, 1998). Even pulsed siltation may lead to a severe loss of seagrass cover, which subsequently requires a long time to return to pre-event levels (Talbot *et al.*, 1990).

Research on the effects of siltation on seagrass ecosystems is still limited, particularly in view of the magnitude of the problem in many regions. Research across South-East Asia has documented a progressive loss of species and seagrass biomass along siltation gradients (Terrados *et al.*, 1998; Fig. 7.9), involving a consistent loss of species according to the sequence: *Syringodium isoetifolium* → *Cymodocea rotundata* → *Thalassia hemprichii* → *Cymodocea serrulata* → *Halodule uninervis* → *Halophila ovalis* → *Enhalus acoroides*. Multispecific seagrass assemblages were confined to sediments with silt contents < 15%, and only *Enhalus acoroides* was able to survive, albeit confined to shallow waters, at highly silted sites. Sediments with silt contents > 20% lacked any seagrass cover. The species that tolerates siltation best, *Enhalus acoroides*, is the largest of the seagrass species, having leaves up to 2 m long. This makes it possible to bring the leaves close to the water surface, thereby avoiding shading by suspended particles in the water column (Vermaat *et al.*, 1997). Species capable of substantial vertical growth, notably *Cymodocea serrulata* (Chapter 2), which are able to raise their leaves well above the sediment level, are only moderately disturbed when experimentally buried. In contrast, *Thalassia hemprichii*, the dominant species in most South-East Asian assemblages, is particularly sensitive to experimental burial, and experiences high mortality (Duarte *et al.*, 1997). Hence, the ranking of species loss with increasing siltation is consistent with predictions derived from the architecture of the regional seagrass flora and their tolerance to experimental burial, and is also consistent with their tolerance to experimental shading (Bach *et al.*, 1998).

7.5.3 *Organic loading of sediments*

As discussed above, eutrophication of coastal waters coincides with increased organic inputs to the sediment, a process that may have a negative impact on seagrasses (Fig. 7.10). Decline of seagrasses has also been linked to the more localized organic loading of the sediment coinciding with aquaculture activities. Fish farming increases pollution of the water due to feed wastage, fish excretion and faecal production (Wu, 1995). Moreover, the sediment underneath the cages is enriched with organic matter due to sedimentation of organic particles. Respiration by the benthic community of fish-farming-influenced sediments may be an order of magnitude higher than in control situations (Holmer & Kristensen, 1992). Cage-farming of gilthead sea bream and sea bass in the Mediterranean has led to serious degradation of *Posidonia oceanica* and *Cymodocea nodosa* meadows in the direct environment, and even complete disappearence beneath the cages (Delgado *et al.*, 1997; Pergent *et al.*, 1999). Most probably such meadow regressions are due to the combined effect of reduced water quality and the organic loading of the sediment. An added reason for concern in the case of organic loading of sediments is that its impact may be long lasting, even if water quality has improved again. This is suggested by the observation that decline of *Posidonia oceanica* in an area of fish culture plants continued for years after closure of the facilities (Delgado *et al.*, 1999).

Fig. 7.10. Catastrophic loss of a *Posidonia oceanica* meadow in south-east Spain. Two years after the operation of a sewage outlet, only the rhizome mattes remained. (Photograph by C.M. Duarte.)

The negative effects of excessive inputs of organic matter on seagrass growth may not just occur as a consequence of human activities. It is not impossible that highly productive seagrass meadows may slowly 'poison' themselves through the input of their own detritus and through the trapping of suspended particles. A highly productive mixed Philippine meadow, for instance, was found to have the most negative sediment redox conditions and the highest sediment sulphide concentrations yet described for any one seagrass meadow (Terrados *et al.*, 1999). Such a meadow may be on the verge of catastrophic mortality. The seasonality of temperate seagrasses may help prevent excessive accumulation of organic material, which may be washed out during winter, when high wave and current energy and low seagrass cover are conducive to sediment resuspension and export of materials. Tropical meadows, however, are able to sustain high production and cover year-round, and may thus be more susceptible to self-poisoning. Disturbance may be necessary for the long-term maintenance of highly productive tropical seagrass meadows, preventing sediment conditions from deteriorating until they impair seagrass growth. Such disturbance may be provided by small-scale events, such as that associated with the burrowing activity of shrimps and crabs, or by large-scale events, such as hurricanes and cyclones.

7.5.4 Toxic chemicals

Human activities lead to pollution of the seas with many substances, some of which are directly toxic to living organisms. Large-scale impacts on seagrasses of potentially toxic pollutants other than oil have not been identified with certainty so far. Oil spills are one of the most conspicuous forms of chemical pollution of the sea. Apart from direct suffocation risks from a covering layer of oil, seagrasses may potentially be damaged by the more soluble oil components that contaminate the seawater. Various degrees of injury to seagrass have been observed after oil spills, from plant mortality and declines in plant cover, to virtually no losses at all (Short & Wyllie-Echeverria, 1996). Generally, the marine communities most vulnerable to oils spills are those in the intertidal range; the risks decrease with increasing depth (Zieman *et al.*, 1984). Subtidal seagrasses fit this pattern as they have been shown to survive serious oil spills (Den Hartog & Jacobs, 1980; Kenworthy *et al.*, 1993; Dean *et al.*, 1998). Seagrass plants also appear to be less sensitive than associated benthic invertebrates (Den Hartog & Jacobs, 1980; Jewett & Dean, 1997, cited in Dean *et al.*, 1998).

7.5.5 Other physico-chemical changes

Human activities may negatively affect seagrass systems by inputs of dissolved and particulate matter to the water column and the sediment, but they may lead to unfavourable environmental conditions in other ways as well. The cooling water of electric power plants, for instance, may alter thermal conditions in the water where discharge takes place, and thermal effluents have been pinpointed as a cause of meadow destruction (Zieman & Wood, 1975). Coastal engineering may also change the environment for seagrasses in a variety of ways. The many dike and dam building projects in the Netherlands this century illustrate the negative impacts such activities may have. In the northern part of the Netherlands the Zuiderzee, an inland sea, was closed off by a 30 km dike in 1932. This changed the hydrodynamics in the adjacent Wadden Sea. The mean tidal range abruptly increased due to the closure, and mean current velocities were raised dramatically at some places (up to three-fold). Re-establishment of eelgrass after the wasting disease in the early

1930s was impossible under these new dynamic conditions (De Jonge & De Jong, 1992). In the south-western part of the country, several tributaries of the Rhine and Meuse rivers were closed off from the sea in order to reduce the risks of inundations during storm floods. One estuary, the Grevelingen, became a saline lake with water that is slowly replenished by restricted entrance of full seawater. Another, the Oosterschelde, was transformed into an arm of the North Sea, without riverine influence. Water transparencies have been excellent for seagrass growth since the completion of the dams in the 1970s and 1980s, yet *Zostera marina* has declined in both estuaries; currently, populations are on the verge of extinction. The loss of the estuarine character of the water appears to have been conducive to these losses. Long-term observations showed a strong positive correlation between dissolved silicon levels in the water and the average percentage cover of eelgrass in Lake Grevelingen (Herman *et al.*, 1996). River water is the main source of dissolved silicon in the marine environment (Tréguer *et al.*, 1995), and the cessation of river inflow thus will have reduced inputs of dissolved silicon. Subsequent experimental observations have not born out a direct positive effect of dissolved silicon on eelgrass growth, suggesting that another, yet unknown, factor covarying with water column silicon levels is important for growth (Kamermans *et al.*, 1999). The increase in water salinity in the Grevelingen and the Oosterschelde, after diversion of the river flow, will have had a negative impact on eelgrass, as growth studies indicate (Kamermans *et al.*, 1999; Van Katwijk *et al.*, 1999). The cumulated evidence suggests that eelgrass populations in the south-west Netherlands were adapted to estuarine conditions (which include moderate salinities), and that the cessation of river inflow led to less favourable conditions, including higher salinity, for the plants.

Decreased freshwater inputs, due to human activities, have also been linked to the changes of the seagrass communities in Florida Bay. Groundwater flow into the bay decreased as a result of extraction of water from the region's aquifers: whereas in 1900 no groundwater was pumped up, more than $2\,500\,000$ m^3 was pumped up daily in the 1990s, causing large-scale decline of mangroves in the region (Tack & Polk, 1999). Canal construction, which diverted drainage water, further contributed to a diminished freshwater inflow in Florida Bay. The decreased freshwater inflow, coinciding with a lower frequency of hurricane-induced disturbance of the system, probably were the major factors that caused a transition in the composition of the seagrass vegetation, from heterogeneous communities of *Halodule wrightii* and *Thalassia testudinum* towards monospecific meadows of the latter species in the 1960s and 1970s (Zieman *et al.*, 1989). Since 1987, thousands of hectares of these *Thalassia* meadows have been lost (Robblee *et al.*, 1991). The causes of this decline are not exactly known, but the self-poisoning effect mentioned above, associated with the formation of very dense *Thalassia* meadows that accumulated large amounts of leaf litter, may have played a role in this decline (Carlson *et al.*, 1994).

7.5.6 Mechanical damage

Fishing practices that disturb bottom sediments may result in severe local reductions of seagrass cover, as shoots and rhizomes are damaged, or completely removed from the substratum by trawling or dredging gear. In the Netherlands, the mussel and cockle fisheries are held partly responsible for declines in eelgrass cover and also for the failure of seagrasses to re-establish in some areas (De Jonge & De Jong, 1992), and extensive damage has also observed elsewhere (Ardizzone & Pelusi, 1984; Fonseca *et al.*, 1984). Dredging for maintenance of shipping channels similarly will have a direct mechanical impact on seagrass meadows, although such losses may be small relative to damage caused by elevated water turbidity (Onuf, 1994). Where small boats are numerous, the cumulative effect of numerous boot moorings and propeller scars may result in considerable loss of vegetation (Walker *et al.*, 1989; Creed & Amado Filho, 1999).

7.5.7 Invasions with exotic species

Invasions of seagrass systems with non-native species have not yet been reported to cause large-scale declines, but such an event is not unlikely. Proliferation of the tropical green alga *Caulerpa taxifolia*, first noted in the Mediterranean in 1984 (Meinesz & Hesse, 1991), is considered as a potential danger to the region's seagrass meadows. The exact way in which this alga reached the Mediterranean, e.g. either by accidental release from aquaria or by migration into the Mediterranean basin through the Suez Canal (Chisholm *et al.*, 1995), remains to be elucidated. The alga is able to grow on most substrata, may spread rapidly, and is able to form very dense meadows to depths of 20 m. It grows well on the root mat of *Posidonia oceanica*, and regression symptoms have been observed in some meadows (De Villèle & Verlaque, 1995). Initial fears for displacement of the region's seagrass meadows by *Caulerpa* have now been reduced somewhat by observations suggesting that only in polluted environments where seagrass vitality is poor, or in sparse meadows, is the alga a successful colonizer (De Villèle & Verlaque, 1995; Chisholm *et al.*, 1997).

7.5.8 Seagrasses and global change

Long-term observations on seagrass meadows are quite rare, and this is one of the reasons that the effect of climate changes over time scales of years and decades is not well documented. Moreover, it is difficult to ascribe changes in areal distribution to climate effects with certainty. The frequent simultaneous occurrence of other environmental changes (e.g. changes in water quality induced by human activities) restricts this possibility. Long-term observations of *Zostera marina* beds in an oceanic setting without pronounced human influence showed clear fluctuations in areal distribution over the 1932–1992 survey period. The gradual decline noticed during the 1980s and 1990s were related to elevated sea-surface temperatures observed since the 1980s (Glémarec *et al.*, 1997). Such observations suggest that expansion and regression of seagrass meadows may follow climate changes, and that widespread changes in the areal distribution of seagrasses as a result of current global change phenomena are not unlikely.

In addition to observational evidence, growing knowledge on physiological ecology of seagrasses is being used to forecast their likely response to global change (Brouns, 1994; Beer & Koch, 1996; Short & Neckles, 1999). Climate change is thought to influence seagrass stands primarily through the predicted rise in sea level (and associated coastal erosion), the increased partial pressure of CO_2 in seawater, and the elevated seawater temperature.

The predicted rise of sea level under the present conditions is about 0.6 cm per year (Brouns, 1994), which is within the rate of vertical growth of most seagrass species (but higher than some, such as *Posidonia oceanica*, cf. Chapter 2). Hence, seagrass canopies may remain at relatively similar depths for long periods of time, provided they can also trap materials to elevate the sediment surface, such as observed in the reef-building *Posidonia oceanica*. The rise in sea level will cause a greater inland penetration of seawater in estuaries, and this may result in the displacement of less salt-tolerant submersed macrophyte species by seagrasses. Sea-level rise will also lead to coastline regression and sediment erosion, expected to be about 6 m per year (applying Bruun's rule, i.e. 10 m recession per cm increase in sea level; Bruun, 1962). Seagrasses are unable to migrate at these rates and, therefore, will probably experience widespread losses due to sediment erosion. Sediment erosion may be aggravated in areas where coastlines are heavily occupied by human constructions, as is becoming evident already in the Mediterranean (Marbà *et al.*, 1996). Such constructions may contribute to the destabilization of the marine sedimentary environment and cause additional seagrass loss.

The increase in atmospheric CO_2 levels has a rather complex impact on the dissolved inorganic carbon sources available for seagrass photosynthesis (CO_2 and HCO_3^-, see section 4.3.1), as the increase in dissolved CO_2 in seawater that follows an enhanced atmospheric CO_2 causes interacting shifts in the equilibrium concentrations of the dissolved inorganic carbon species and in pH (Goudriaan, 1993; Short & Neckles, 1996). The overall consequence of rising atmospheric CO_2 will be an increase in both the actual concentration of dissolved CO_2 and the relative proportion of CO_2 to HCO_3^-. This could, in turn, enhance the productivity of dense seagrass beds, which are likely to experience carbon limitation (see section 4.3.2). In addition, the depth range of seagrasses may be extended, as the higher photosynthetic rates at elevated dissolved CO_2 levels would be conducive to a positive carbon balance at greater depths. These positive effects on seagrass growth, however, could be at least partially offset by stimulated growth of epiphytic algae on the leaf surfaces. These might also positively respond to increased availability of CO_2, preventing or diminishing the productivity increase expected in seagrasses. Another uncertainty concerns the physiological adjustments to elevated dissolved CO_2 levels in seagrasses, which could include a general down-regulation of photosynthesis, as has been observed in other submerged angiosperms (Madsen *et al.*, 1996).

Scenarios for temperature rise in the next century are continually adapted, but the crucial point remains that global mean temperatures are expected to rise considerably faster in the next century than they have done in the past century. As is the case for the impact of increased atmospheric levels of CO_2, predictions of the effect of increased seawater temperatures on seagrasses are still very speculative, and much more research is required to allow the formulation of more specific predictions. Obviously, the direct effects of increased temperatures will depend on individual species' thermal tolerances and on the specific temperature dependency of the many processes that determine growth and reproduction. For species living near their upper temperature limit, a further increase in temperature may be fatal. In less critical environments, a rise in seawater temperature may have an impact on processes like photosynthesis, respiration, nutrient uptake, flowering and seed germination (Short & Neckles, 1999). Such effects will have consequences for the vigour and competitive ability of populations. It is therefore likely that the species-dependent impact of temperature increase will alter seagrass distribution and abundance in the long term.

Increased ultraviolet (UV) radiation, finally, may also be a reason of concern for the vitality of seagrasses. Reductions in stratospheric ozone by anthropogenic agents allow more UV radiation to reach the earth's surface, particularly in the 280–315 nm band (UV-B), and to a lesser extent also in the 315–400 nm band (UV-A). Solar radiation is rapidly absorbed and scattered as it passes through the water column (section 4.2.1), but these parts of the radiation spectrum are close to the wavelengths that are the least attenuated in pure water. Hence, UV-B may penetrate to depths greater than 30 m in clear oceanic waters (Smith & Baker, 1979). Extensive research on terrestrial plants has shown that UV-B has various harmful effects, including a reduction in photosynthetic capacity (Teramura & Sullivan, 1994). The relative strong penetration of UV-B in water implies that increased levels of UV-B may also affect the functioning of seagrasses, particularly those growing at shallow depths. Currently, only a few studies have addressed the effect of increased UV radiation on seagrasses. These studies indicate that seagrass species differ considerably in their ability to tolerate such increases. In the most comprehensive study to date, it was found that *Halophila ovalis* and *Halodule uninervis* were relatively sensitive species, showing a conspicuous decrease in photosynthetic efficiency and chloroplast density, and only a small increase in UV-blocking pigments, whereas *Zostera capricorni*, *Cymodocea serrulata* and *Syringodium isoetifolium* were more UV-B tolerant (Dawson & Dennison, 1996). The thickness of the leaves appeared to be an important feature in the sensitivity of the species, with thicker leaves providing greater tolerance to increased UV-B.

Among the various threats that the globally changing environment holds for seagrasses, increased UV-B radiation does not rank as the most important one. Much of the world's seagrass vegetation is not reached by harmful levels of UV-B; furthermore, model simulations of stratospheric ozone indicate that recovery of the ozone layer will gradually take place over the next 50 years, following the phasing out of chlorine–fluorine hydrocarbon compounds (Madronich *et al.*, 1995).

7.6 Monitoring seagrass meadows

Repeated determination of the areal cover and density of seagrasses is a well-tried method to assess the status of the vegetation and to detect declines. Direct observation, from a boat, snorkling, or by scuba divers, of the distribution of the meadows is a rather time-consuming procedure certainly if large areas have to be mapped, and samples for biomass or shoot density measurements additionally have to be collected. However, the method is straightforward and effective in detecting declines. Particularly important for the credibility and the usefulness of the observations, e.g. in communicating them to regional and state regulatory agencies, is the fact that changes in cover are exactly quantified, and that spatial patterns in the distributional changes are documented (e.g. Nienhuis *et al.*, 1996). Instead of measuring the entire meadow area periodically, it also possible to position some permanent transects in the seagrass vegetation at sites that are representative for larger areas of the meadow, and record any changes in these transects (Kirkman, 1996). Particularly when very large areas with seagrasses need to be monitored, direct observation as described above soon becomes impractical. In such instances, satellite or airborne remote sensing is a possible alternative. These methods are only feasible in clear water areas, where seagrass meadows are recognizable by reflective properties that distinguish them from other underwater features. Ground surveys remain important when using remote sensing techniques to verify the presence of seagrass meadows, and to eliminate the possibility that underwater features such as beds of macroalgae, reefs, or detritus banks are erroneously identified as seagrass meadows.

The detection of declines of seagrass meadows, although important because it gives information on the health status of the meadows, and has a signal function indicating environmental deterioration, is rather unsatisfactory for managers, for experience on seagrass losses shows that once losses become apparent it is probably too late to counteract the disturbance. Ideally, monitoring techniques should be able to forecast losses before these occur. As described earlier, seagrass decline requires reduced shoot recruitment, increased shoot mortality or both, conducive to a negative net rate of population change. Reconstruction techniques to estimate the age of seagrass shoots and, from this, the net rate of population change have been used to issue forecasts on the likely change of seagrass populations in the north-west Mediterranean (Marbà *et al.*, 1996) and Florida Bay (Durako, 1994). Although subject to some controversy (Jensen *et al.*, 1996; Kaldy *et al.*, 1999), these forecasts have been shown to reproduce the trend in seagrass populations at least for the following year (Durako & Duarte, 1997). However, verification of actual population changes in some of the Mediterranean meadows (Marbà *et al.*, 1996) showed that, whenever a tendency towards decline was predicted, realized rates exceeded the expected ones, because seagrass decline triggers a number of negative feedback effects that accelerate the loss (see section 7.5.1). Consequently, even these attempts to anticipate seagrass loss may be insufficient as once a negative imbalance between seagrass recruitment and mortality is detected, it is, again, too late to prevent the damage. Future research must address this problem and lead to the development of early warning techniques. Because seagrass decline ultimately depends on meristematic vigour, direct consideration of meristematic activity probably is a promising approach.

Analysis of chlorophyll *a* fluorescence has been used as a tool for stress detection in terrestrial plants for over a decade. In recent years, this approach is being explored in seagrasses. The method is based on the fact that quenching of chlorophyll following absorption of light quanta not only includes transfer of energy to the reaction centres of the two photosystems to drive the primary photochemical reactions, but that for a minor part deactivation of the light-excited chlorophyll molecules also occurs by fluorescence emission (Krause & Weiss, 1991). In dark-adapted plants, maximal fluorescence (F_m) occurs when all reaction centres in photosystem II are reduced, or closed, by a preceding short dose of saturating light. The initial fluorescence (F_0) is measured when the maximum number of reaction centres are available for photon reduction; hence, this level is measured in dark-adapted plants that have not been briefly exposed to a light dose. The difference between F_m and F_0 is called F_v, the variable fluorescence. Similar parameters can be measured in light-adapted plants, with F_m' being the maximal fluorescence, F the fluorescence in the ambient light, and ΔF being the difference between F_m' and F. Changes in the F_v/F_m and $\Delta F/F_m'$ ratios can be indicative of stress. Laboratory experiments show that in *Halophila ovalis*, high irradiance stress, osmotic stress, thermal stress and exposure to petro-chemicals all lead to decreases in the F_v/F_m or $\Delta F/F_m'$ ratios (Ralph & Burchett, 1995; 1998; Ralph, 1998a,b). Increased UV-B reduced F_v/F_m ratios in several seagrass species (Dawson & Dennison, 1996), whereas *Zostera marina* growth at elevated salinity also coincided with a decrease in the F_v/F_m ratio (Kamermans *et al.*, 1999). An example of such changes is presented in Fig. 7.11, which shows that in experiments investigating

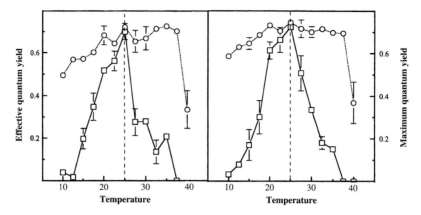

Fig. 7.11. *Halophila ovalis*. The response of the $\Delta F/F_\mathrm{m}'$ (effective quantum yield) and the $F_\mathrm{v}/F_\mathrm{m}$ (maximum quantum yield) ratio to a range of temperatures, after exposure for 5 h (○) or 96 h (□). Means and 95% confidence intervals (n=4). Units of the ratios are relative. (Ralph, 1998a.)

the effects of thermal stress in *Halophila ovalis*, both the $F_\mathrm{v}/F_\mathrm{m}$ and the $\Delta F/F_\mathrm{m}'$ ratio show significant decreases relative to the values measured at 25 °C after 96 h acclimation at lower and higher temperatures (Ralph, 1998a). A draw-back of the method presently seems that, unlike in the example of Fig. 7.11, other types of stresses may induce ratio shifts only in rather extreme situations (Ralph, 1998b). The value of this method for detecting stress in seagrasses, furthermore, has yet to be proven in field situations.

7.7 Restoration of seagrass meadows

The widespread loss of seagrass meadows, coinciding with a growing knowledge and awareness of the resource value of these systems, has led to increasing attention to restoring meadows in areas where they were lost, or to create them in areas that appear suitable for growth, in an effort to compensate for losses elsewhere. In the following, we will loosely use the term 'restoration' for both activities, although, strictly speaking, it only applies to re-establishment of a meadow at a former growth site. In many countries protection of seagrass meadows is now governmental policy. In the US, current legislation even requires that, in the case that losses due to specific local developmental activities cannot be prevented, the lost area must be replaced by habitat with equivalent functional values (Davis & Short, 1997). Not surprisingly, it is in the US that most experience has been gained with the restoration of seagrass meadows. For a compilation of US data and extensive guidelines, the interested reader is referred to Fonseca *et al.* (1998). In many other countries, including Australia and several European countries, restoration of seagrass meadows is currently an issue receiving considerable attention as well (Genot *et al.*, 1994; Christensen *et al.*, 1995; Van Katwijk *et al.*, 1998; Balestri *et al.*, 1998; Paling *et al.*, 1998).

Creating a seagrass meadow at a given site, until now, implies the transplantation of plant material from existing meadows to the new location. The plant material that is used typically consists either of shoots with roots and rhizomes in the original sediment dug out from the donor site, or of shoots with the roots and rhizomes washed free of sediment. A variety of devices is used to obtain the plugs of seagrass with the associated sediment, and the size of these 'planting units' varies with the method that is applied (Fonseca, 1994; Fonseca *et al.*, 1998; Davis & Short, 1997). In most cases, the planting units have been relatively small, e.g. when taken with metal corers with a diameter of no more than 10 cm, but they can also be of a size that can no longer be handled by an individual person. In a recent transplantation programme in Western Australia, large sods with a surface area of 0.25 m^2 and weights of approximately 350 kg were harvested, which was possible by the use of a newly developed, submarine operating, machine transplanting system (Paling *et al.*, 1998). A disadvantage of using plants with associated sediment is that holes are created in the sediment of the donor site, which, even when they have been filled, are susceptible to erosion. When plants are carefully freed from the sediment at the donor site, to obtain shoots with bare rhizome and roots for transplantation, the damage to the donor sediment is much less, and this is an argument to use plants without associated sediment. Applying this method, however, some damage to the subterranean parts of the donor plants is inevitable. Such damage may weaken the plants, and may have consequences for the transplantation success. In addition, the isolated plants are often easily lost from the site by hydrodynamic action. When shoots with bare roots and rhizomes are used for transplantation, they are therefore often anchored in the recipient sediment by some means, e.g. by U-shaped staples, to provide stabilization of the transplants until they are firmly rooted in the new environment.

The simplest way to restore seagrass meadows would seem to be the broadcasting of seeds. Yet this procedure is much less frequently applied than transplantation of intact plants. The reason for this varies among species. In some cases, seeds are difficult or impossible to obtain in reasonable quantities. In other cases, seed viability and seedling survival are very low or highly unpredictable, making this method rather unsuitable when time for restoration activities is limited, and repeated efforts for this reason are not possible.

Whatever the nature of the transplanted material, it currently requires the collection of material from existing meadows, which are potentially negatively affected by this activity. In the case of a slow-spreading species such as *Thalassia testudinum*, harvesting of parts of the vegetation for transplantation purposes may leave a bare patch for years (Fonseca, 1994). Even the collection of seeds from a population must be done prudently, as natural seedling recruitment may be necessary for persistent meadow vitality. Apart from these considerations, transplantation is not to be undertaken lightly, as the effort needed to establish a new meadow of an even moderate size can be spectacular. For example, to compensate for the impacts of a harbour expansion project on *Zostera marina* beds, 2.5 ha of new meadows were planted in the Great Bay Estuary, New Hampshire. This required the collection of 250 000 3–5 cm rhizome pieces with associated shoots and roots, each individually harvested from the donor site; after collection each of these plant pieces was transplanted by scuba divers to the intertidal and subtidal sites selected for the new meadows (Davis & Short, 1997).

Transplantations have been carried out with at least six different seagrass species, the three most important ones being *Zostera marina*, *Halodule wrightii* and *Thalassia testudinum* (Fonseca *et al.*, 1998). The survival of the planting units on the transplant sites is highly variable among the various projects. Many factors determine the success, but the proper choice of the transplant site certainly is crucial. Transplantation to sites with environmental conditions that approach those at the donor site as much as possible is the most obvious way to improve the survival chances of the transplants. Furthermore, transplant sites should preferably have supported seagrass in the past (Fonseca, 1994). Succesful restoration of a meadow at a site where seagrass disappeared because of deterioration of environmental conditions, of course, can only be expected if the cause of the initial decline has been abated. Assessing transplantation success requires monitoring of the site, preferably for several years. Survival of the planting units, area covered by the expanding planting units, and shoot density are among the parameters that can be measured for this purpose. A recent compilation of data on

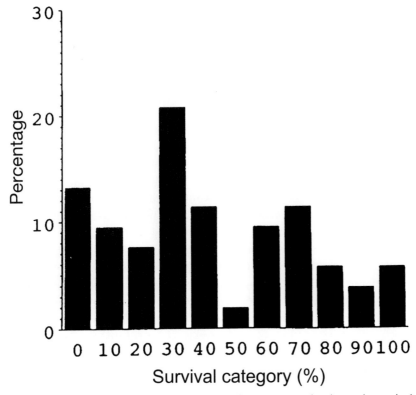

Fig. 7.12. Frequency distribution of percentage planting unit survival in 53 seagrass transplantation projects carried out in the USA. *Y*-axis gives percentage of the survival values falling in the percentage survival categories on the *x*-axis. (Fonseca *et al.*, 1998.)

planting unit survival, derived from 53 reports published in the US, shows a median percentage planting unit survival of 35%, and a mean percentage of 42% (Fig. 7.12; Fonseca *et al.*, 1998). It is thus clear that generally many of the planting units are lost, a fact that must be incorporated in the planning of any transplantation project.

A seagrass ecosystem is more than just vegetation, but, in natural meadows, contains an abundance of animals as well (Chapter 6). Several studies indicate that planted seagrass beds are colonized by many animals in the course of time, and the transplanted beds thus gradually may develop into the diverse systems of natural beds (Homziak *et al.*, 1982; McLaughlin *et al.*, 1983; Fonseca *et al.*, 1990; 1996). The composition and abundance of the faunal community, however, depends on the successful development of the bed. This can be gathered from observations of Fonseca *et al.* (1990), who found that in an area where *Zostera marina* transplants performed poorly, the fish and decapod fauna differed in composition and had a lower abundance in comparison with natural beds, whereas in an area where transplants developed favourably, faunal abundance and composition 1.9 years after transplantation were indistinguishable from natural eelgrass beds. A subsequent study on fish and decapod abundance in planted *Halodule wrightii* and *Syringodium filiforme* beds showed that faunal density displayed an asymptotic relationship with shoot density, and became equal to natural beds already at shoot densities a third of natural shoot densities (Fonseca *et al.*, 1996). Hence, there is no reason to believe that planted seagrass beds, if they persist and grow well, would be inferior to natural beds in terms of faunal abundance and diversity. There is some uncertainty, however, with respect to the genetic diversity of transplanted beds. A study of transplanted eelgrass beds in southern California suggested that genetic diversity of the plants (as determined by allozyme electrophoresis) was reduced compared with natural eelgrass beds in the area (Williams & Davis, 1996). As lack of genetic diversity is expected to make populations more uniformly susceptible to diseases or other disturbances, reduced genetic diversity in transplanted seagrass would tend to diminish the perspectives for long-term survival. Further investigations on this subject are thus needed.

7.8 Epilogue

The preceding discussion portrays seagrass meadows as threatened ecosystems. The many losses already experienced clearly indicate the need for implementing policies that ensure the sustainability of the extant seagrass meadows, as well as the desirability to re-establish those that were already lost. This is, however, a gigantic task, and one that is located at the interface between ecology, socio-economic issues and politics, which will be the setting of decision-making on the planet's ecosystems in the twenty-first century. While it is clear that this and the preceding chapters did not, and were not intended to, address this complex scenario, we consider the capacity to provide sound scientific advice based on a solid understanding of seagrass ecology as a necessary prerequisite for the successful management of these ecosystems. We hope that the reader who has persevered through the book's chapters to find this epilogue does so endowed with a basic understanding of seagrass ecology, from their architecture as plants to the roles they perform in the ecosystem, allowing her or him to identify best management practices that help preserve or recover these meadows, or to identify areas where research is needed to best attain this goal. If this were the case, we would feel amply compensated by our dedication and efforts to write this book under the permanent struggle between keeping the book brief and simple enough to be available to the widest possible audience and the guilty feeling of leaving out so many fascinating details on the ecology of seagrasses.

7.9 References

Agustí, S., Enríquez, S., Christensen, H., Sand-Jensen, K. & Duarte, C.M. (1994). Light harvesting by photosynthetic organisms. *Functional Ecology*, **8**, 273–79.

Ardizzone, G.M. & Pelusi, P. (1984). Yield damage of bottom trawling on *Posidonia oceanica* meadow. In *International Workshop on* Posidonia oceanica *Beds*, ed. C.F. Boudouresque, A.J. de Grissac & J. Olivier, GIS Posidonie publ. **1**, 63–72.

Bach, S.S., Borum, J., Fortes, M.D. & Duarte, C.M. (1998). Species composition and plant performance of mixed seagrass beds along a siltation gradient at Cape Bolinao, The Philippines. *Marine Ecology Progress Series*, **174**, 247–56.

Balestri, E., Piazzi, L. & Cinelli, F. (1998). Survival and growth of transplanted and natural seedlings of *Posidonia oceanica* (L.) Delile in a damaged coastal area. *Journal of Experimental Marine Biology and Ecology*, **228**, 209–25.

Beer, S. & Koch, E. (1996). Photosynthesis of seagrasses vs. marine macroalgae in globally changing CO_2 environments. *Marine Ecology Progress Series*, **141**, 199–204.

Bethoux, J.P. & Copin-Montégut., G. (1986). Biological fixation of atmospheric nitrogen in the Mediterranean Sea. *Limnology and Oceanography*, **31**, 1353–8.

Borum, J. (1985). Development of epiphytic communities on eelgrass (*Zostera marina*) along a nutrient gradient in a Danish estuary. *Marine Biology*, **87**, 211–18.

Borum, J. (1996). Shallow waters and land/sea boundaries. In *Eutrophication in Coastal Marine Ecosystems*, ed. K. Richardson & B.B. Jorgensen, pp. 179–203. Washington: American Geophysical Union.

Borum, J. & Sand-Jensen, K. (1996). Is total primary production in shallow coastal marine waters stimulated by nitrogen loading? *Oikos*, **76**, 406–10.

Brøndegaard, V.J. (1987). *Folk og Flora 1*. København: Dansk etnobotanik.

Brouns, J.J. (1994). Seagrasses and climate change. In *Impacts of Climate Change on Ecosystems and Species: marine and coastal systems*, ed. J.C. Pernetta, R. Leemans, D. Elder & S. Humphrey, pp. 59–71. Gland: IUCN.

Bruun, P. (1962). Sea level rise as a cause of shore erosion. *Journal of the Waterways and Harbours Division*, Proceedings of the American Society of Civil Engineers, **88**, 117–30.

Burkholder, J.M., Mason, K.H. & Glasgow, Jr., H.B. (1992). Water-column nitrate enrichment promotes decline of eelgrass *Zostera marina*: evidence from seasonal mesocosm experiments. *Marine Ecology Progress Series*, **81**, 163–78.

Burkholder, J.M., Glascow, H.B. & Cooke, J.E. (1994). Comparative effects of water column nitrate enrichment on eelgrass *Zostera marina*, shoalgrass *Halodule wrightii*, and widgeongrass *Ruppia marina*. *Marine Ecology Progress Series*, **105**, 121–38.

Cambridge, M.L. & McComb, A.J. (1984). The loss of seagrasses in Cockburn Sound, Western Australia. I. The time course and magnitude of seagrass decline in relation to industrial development. *Aquatic Botany*, **20**, 229–43.

Carlson, P.R., Yabro, L.A., Barber, T.R. (1994). Relationship of sediment sulfide to mortality of *Thalassia testudinum* in Florida Bay. *Bulletin of Marine Science*, **54**, 733–46.

Cebrián, J. & Duarte, C.M. (1994). The dependence of herbivory on growth rate in natural plant communities. *Functional Ecology*, **8**, 518–25.

Chisholm, J.R.M., Jaubert, J.M. & Giaccone, G. (1995). *Caulerpa taxifolia* in the northwest Mediterranean: introduced species or migrant from the Red Sea? *Comptes Rendus de l'Academie des Sciences, Paris, Sciences de la Vie*, **318**, 1219–26.

Chisholm, J.R.M., Fernex, F.E., Mathieu, D. & Jaubert, J.M. (1997). Wastewater discharge, seagrass decline and algal proliferation on the Côte d'Azur. *Marine Pollution Bulletin*, **34**, 78–84.

Christensen, P.B., Sortkjær, O. & McGlathery, K.J. (1995). *Transplantation of Eelgrass*. Silkeborg, Denmark: National Environmental Research Institute.

Christiansen, C., Christoffersen, H., Dalsgaard, J. & Norberg, R. (1981). Coastal and nearshore changes correlated with die-back in eelgrass (*Zostera marina* L.). *Sedimentary Geology*, **28**, 163–73.

Coleman, V.L. & Burkholder, J.M. (1995). Response of microalgal epiphyte communities to nitrate enrichment in an eelgrass (*Zostera marina*) meadow. *Journal of Phycology*, **31**, 36–43.

Costanza, R., d'Arge, R., De Groot, R., Fraber, S., Grasso, M., Hannon, B., Limburg, K., Naeem, S., O'Neill, R.V., Paruelo, J., Raskin, R.G., Sutton, P., Van den Belt, M. (1997). The value of the world's ecosystem services and natural capital. *Nature*, **387**, 253–60.

Creed, J.C. & Amado Filho, G.M. (1999). Disturbance and recovery of the macroflora of a seagrass (*Halodule wrightii* Ascherson) meadow in the Abrolhos Marine National Park, Brazil: an experimental evaluation of anchor damage. *Journal of Experimental Marine Biology and Ecology*, **235**, 285–306.

Davis, R.C. & Short, F.T. (1997). Restoring eelgrass, *Zostera marina* L., habitat using a new transplantation technique: the horizontal rhizome method. *Aquatic Botany*, **59**, 1–15.

Dawson, S.P. & Dennison, W.C. (1996). Effects of ultraviolet and photosynthetically active radiation on five seagrass species. *Marine Biology*, **125**, 629–38.

Dean, T.A., Stekoll, M.S., Jewett, S.C., Smith, R.O. & Hose, J.E. (1998). Eelgrass (*Zostera marina* L.) in Prince William Sound, Alaska: effects of the Exxon Valdez oil spill. *Marine Pollution Bulletin*, **36**, 201–10.

De Jonge, V.N. & De Jong, D.J. (1992). Role of tide, light and fisheries in the decline of *Zostera marina* L. in the Dutch Wadden Sea. *Netherlands Institute for Sea Research Publication Series*, **20**, 161–76.

Delgado, O., Grau, A., Pou, S., Riera, F., Massuti, C., Zabala, M. & Ballesteros,

E. (1997). Seagrass regression caused by fish cultures in Fornells Bay, Menorca, western Mediterranean. *Oceanologica Acta*, **20**, 557–63.

Delgado, O., Ruiz, J., Pérez, M., Romero, J. & Ballesteros, E. (1999). Effects of fish farming on seagrass (*Posidonia oceanica*) in a Mediterranean bay: seagrass decline after organic loading cessation. *Oceanologica Acta*, **22**, 109–17.

Den Hartog, C. & Jacobs, R.P.W.M. (1980). Effects of the Amoco Cadiz oil spill on an eelgrass community at Roscoff (France) with special reference to the mobile benthic fauna. *Helgolaender Meeresuntersuchungen*, **33**, 182–91.

Den Hartog, C. (1987). 'Wasting disease' and other dynamic phenomena in *Zostera* beds. *Aquatic Botany*, **27**, 3–14.

Den Hartog, C. (1994). Suffocation of a littoral *Zostera* bed by *Enteromorpha radiata*. *Aquatic Botany*, **47**, 21–8.

Dennison, W.C., Orth, R.J., Moore, K.A., Stevenson, J.C., Carter, V., Kollar, S., Bergstrom, P.W. & Batiuk, R. (1993). Assessing water quality with submersed aquatic vegetation. *Bioscience*, **43**, 86–91.

De Villèle, X. & Verlaque, M. (1995). Changes and degradation in a *Posidonia oceanica* bed invaded by the introduced tropical alga *Caulerpa taxifolia* in the North Western Mediterranean. *Botanica Marina*, **38**, 79–87.

Downing, J.A., Osenberg, C.W. & Sarnelle, O. (1999). Meta-analysis of marine nutrient-enrichment experiments: variation in the magnitude of nutrient limitation. *Ecology*, **80**, 1157–67.

Duarte, C.M. (1992). Nutrient concentrations of aquatic plants: patterns across species. *Limnology and Oceanography*, **37**, 882–9.

Duarte, C.M. (1995). Submerged aquatic vegetation in relation to different nutrient regimes. *Ophelia*, **41**, 87–112.

Duarte, C.M. & Cebrián, J. (1996). The fate of marine autotrophic production. *Limnology and Oceanography*, **41**, 1758–66.

Duarte, C.M., Terrados, J., Agawin, N.S.W., Fortes, M.D., Bach, S. & Kenworthy, W.J. (1997). Response of a mixed Philippine seagrass meadow to experimental burial. *Marine Ecology Progress Series*, **147**, 285–94.

Durako, M.J. (1994). Seagrass die-off in Florida Bay: changes in shoot demographic characteristics and population dynamics. *Marine Ecology Progress Series*, **110**, 59–66.

Durako, M.J. & Duarte, C.M. (1997). On the use of reconstructive aging techniques for assessing seagrass demography: a critique of the model test of Jensen *et al.* (1996*). Marine Ecology Progress Series*, **146**, 297–303.

Durako, M.J. & Kuss, K.M. (1994). Effects of *Labyrinthula* infection on the photosynthetic capacity of *Thalassia testudinum*. *Bulletin of Marine Science*, **54**, 727–32.

Enríquez, S., Nielsen, S.L., Duarte, C.M. & Sand-Jensen, K. (1996). Broad-scale comparison of photosynthetic rates across phototrophic organisms. *Oecologia* (Berlin) **108**, 197–206.

Felger, R.S., Moser, M.B. & Moser, E.W. (1980). Seagrasses in Seri Indian culture. In *Handbook of Seagrass Biology: an ecosystem perspective*, ed. R.C. Phillips & C.P. McRoy, pp. 260–76. New York: Garland STPM Press.

Font-Quer, P. (1980). *Plantas Medicinales*. Barcelona: Labor.

Fonseca, M.S. (1989). Sediment stabilization by *Halophila decipiens* in comparison to other seagrasses. *Estuarine and Coastal Shelf Science*, **29**, 501–7.

Fonseca, M.S. (1994). *A Guide to Planting Seagrasses in the Gulf of Mexico*. Texas A & M University, Sea Grant College Program.

Fonseca, M.S., Thayer, G.W. & Chester, A.J. (1984). Impact of scallop harvesting on eelgrass (*Zostera marina*) meadows. Implications for management. *North American Journal of Fisheries Management*, **4**, 286–93.

Fonseca, M.S., Kenworthy, W.J., Colby, D.R., Rittmaster, K.A. & Thayer, G.W. (1990). Comparisons of fauna among natural and transplanted eelgrass *Zostera marina* meadows: criteria for mitigation. *Marine Ecology Progress Series*, **65**, 251–64.

Fonseca, M.S., Meyer, D.L. & Hall, M.O. (1996). Development of planted seagrass beds in Tampa Bay, Florida, USA. II. Faunal components. *Marine Ecology Progress Series*, **132**, 141–56.

Fonseca, M.S., Kenworthy, W.J. & Thayer, G.W. (1998). *Guidelines for the conservation and restoration of seagrasses in the United States and adjacent waters*. NOAA Coastal Ocean Decision Analysis Series No. 12. NOAA Coastal Ocean Office, Silver Spring, MD.

Fortes, M.D. (1988). Mangrove and seagrass beds of East Asia: habitats under stress. *Ambio*, **17**, 207–13.

Genot, I., Caye, G., Meinesz, A. & Orlandini (1994). Role of chlorophyll and carbohydrate contents in survival of *Posidonia oceanica* cuttings transplanted to different depths. *Marine Biology*, **119**, 23–9.

Giesen, W.B.J.T., Van Katwijk, M.M. & Den Hartog, C. (1990). Temperature, salinity, insolation and wasting disease of eelgrass (*Zostera marina* L.) in the Dutch Wadden Sea in the 1930's. *Netherlands Journal of Sea Research*, **25**, 395–404.

Glémarec, M. (1979). Les fluctuations temporelles des peuplements benthiques liées aux fluctuations climatiques. *Oceanologica Acta*, **2**, 365–71.

Glémarec, M., Le Faou, Y. & Cuq, F. (1997). Long-term changes of seagrass beds in the Glenan Archipelago (South Brittany). *Oceanologica Acta*, **20**, 217–27.

Gómez, E.D. (1988). Overview of environmental problems in the East Asian Seas Region. *Ambio*, **17**, 166–69.

Gordon, D.M., Grey, K.A., Chase, S.C. & Simpson, C.J. (1994). Changes to the structure and productivity of a *Posidonia sinuosa* meadow during and after imposed shading. *Aquatic Botany*, **47**, 265–75.

Goudriaan, J. (1993). Interaction of ocean and biosphere in their transient responses to increasing atmospheric CO_2. *Vegetatio*, **104/105**, 329–37.

Harlin, M.M. & Thorne-Miller, B. (1981). Nutrient enrichment of seagrass beds in a Rhode Island coastal lagoon. *Marine Biology*, **65**, 221–9.

Hemminga, M.A. & Nieuwenhuize, J. (1990). Seagrass wrack-induced dune formation on a tropical coast (Banc d'Arguin, Mauritania). *Estuarine, Coastal and Shelf Science*, **31**, 499–502.

Herman, P.M.J. & Scholten, H. (1990). Can suspension feeders stabilise estuarine ecosystems? In *Trophic Relationships in the Marine Environment, Proceedings of the 24th European Marine Biology Symposium*, ed. M. Barnes & R.N. Gibson, pp. 104–16. Aberdeen University Press.

Herman, P.M.J., Hemminga, M.A., Nienhuis, P.H., Verschuure, J.M. & Wessel, E.G.J. (1996). Wax and wane of eelgrass *Zostera marina* and water column silicon levels. *Marine Ecology Progress Series*, **144**, 303–7.

Hine, A.C., Evans, M.W., Davis, R.A., Jr. & Belknap, D.F. (1987). Depositional response to seagrass mortality along a low-energy, barrier-island coast: West-Central Florida. *Journal of Sedimentary Petrology*, **57**, 431–9.

Holmer, M. & Kristensen, E. (1992). Impact of marine fish cage farming on metabolism and sulfate reduction of underlying sediments. *Marine Ecology Progress Series*, **80**, 191–201.

Homziak, J., Fonseca, M.S., Kenworthy, W.J. & Thayer, G.W. (1982). Macrobenthic community structure in a transplanted eelgrass (*Zostera marina*) meadow. *Marine Ecology Progress Series*, **9**, 211–21.

Jensen, S.L., Robbins, B.D. & Bell, S.S. (1996). Predicting population decline: seagrass demographics and the reconstructive method. *Marine Ecology Progress Series*, **136**, 267–76.

Kaldy, J.E., Fowler, N. & Dunton, K.H. (1999). Critical assessment of *Thalassia testudinum* (turtle grass) aging techniques: implications for demographic inferences. *Marine Ecology Progress Series*, **181**, 279–88.

Kamermans, P., Hemminga, M.A. & De Jong, D.J. (1999). Significance of salinity and silicon levels for growth of a formerly estuarine eelgrass (*Zostera marina*) population (Lake Grevelingen). *Marine Biology*, **133**, 527–39.

Kenworthy, W.J., Durako, M.J., Fatemy, S.M.R., Valavi, H. & Thayer, G.W. (1993). Ecology of seagrasses in northeastern Saudi Arabia one year after the Gulf war oil spill. *Marine Pollution Bulletin*, **27**, 213–22.

Kirkman, H. (1996). Baseline and monitoring methods for seagrass meadows. *Journal of Environmental Management*, **47**, 191–201.

Kirkman, H. & Kendrick, G.A. (1997). Ecological significance and commercial harvesting of drifting and beach-cast macro-algae and seagrasses in Australia: a review. *Journal of Applied Phycology*, **9**, 311–26.

Kirkman, H. & Walker, D.I. (1989). Regional studies – Western Australian seagrasses. In *Biology of Seagrasses*, ed. A.W.D Larkum, A.J. McComb & S.A. Shepherd, pp. 157– 181. Amsterdam: Elsevier.

Krause, G.H. & Weis, E. (1991). Chlorophyll fluorescence and photosynthesis: the basics. *Annual Review of Plant Physiology and Plant Biology*, **42**, 313–49.

Larkum, A.W.D. & West, R.J. (1990). Long-term changes of seagrass meadows in Botany Bay, Australia. *Aquatic Botany*, **37**, 55–70.

Lee, S.Y. (1994). Grave threats to seagrass. *Marine Pollution Bulletin*, **28**, 196.

Lin, H.-J., Nixon, S.W., Taylor, D.I., Granger, S.L. & Buckey, B.A. (1996). Responses of epiphytes on eelgrass, *Zostera marina* L., to separate and combined nitrogen and phosphorus enrichment. *Aquatic Botany*, **52**, 243–58.

Madsen, T.V., Maberly, S.C. & Bowes, G. (1996). Photosynthetic acclimation of submersed angiosperms to CO_2 and HCO_3^-. *Aquatic Botany*, **53**, 15–30.

Madronich, S., McKenzie, R.L., Caldwell, M.M. & Björn, L.O. (1995). Changes in ultraviolet radiation reaching the earth's surface. *Ambio*, **24**, 143–52.

Malmer, A. & Grip, H. (1994). Converting tropical rain forest to forest plantation in Sabah, Malaysia. Part II. Effects on nutrient dynamics and net losses in streamwater. *Hydrobiological Processes*, **8**, 195–209.

Marbà, N. & Duarte, C.M. (1997). Interannual changes in seagrass (*Posidonia oceanica*) growth and environmental change in the Mediterranean littoral zone. *Limnology & Oceanography*, **42**, 800–10.

Marbà, N., Duarte, C.M., Cebrián, J., Enríquez, S., Gallegos, M.E., Olesen, B., & Sand-Jensen, K. (1996). Growth and population dynamics of *Posidonia oceanica* on the Spanish Mediterranean coast: elucidating seagrass decline. *Marine Ecology Progress Series*, **137**, 203–13.

McGlathery, K.J. (1995). Nutrient and grazing influences on a subtropical seagrass community. *Marine Ecology Progress Series*, **122**, 239–52.

McLaughlin, P.A., Treat, S.F., Thorhaug, A., Lemaitre, R. (1983). A restored seagrass (*Thalassia*) bed and its animal community. *Environmental Conservation*, **10**, 247–54.

McRoy, C.P. & Helfferich, C. (1980). Applied aspects of seagrasses. In *Handbook of Seagrass Biology: an ecosystem perspective*, ed. R.C. Phillips & C.P. McRoy, pp. 297–343. New York: Garland STPM Press.

Meinesz, A. & Hesse, B. (1991). Introduction et invasion de l'algue tropicale *Caulerpa taxifolia* en Méditerranée nord-occidentale. *Oceanologica Acta*, **14**, 415–26.

Milliman, J.D., & Meade, R.H. (1993). World-wide delivery of river sediment to the oceans. *Journal of Geology*, **91**, 1–21.

Mitchell, A.W., Bramley, R.G.V. & Johnson, A.K.L. (1997). Export of nutrients and suspended sediment during a cyclone-mediated flood event in the Herbert river catchment, Australia. *Marine and Freshwater Research*, **48**, 79–88.

Molinier, R.M. & Picard, J. (1952). Recherches sur les herbiers de phanérogames marines du littoral Méditeranéen français. *Annales de l'Institut Oceanographique*, **29**, 157–234.

Montaño, N.M., Bonifacio, R.S., Rumbaoa, G.O. (1999). Proximate analysis of the flour and starch from *Enhalus acoroides* (L.f.) Royle seeds. *Aquatic Botany*, **65**, 321–5.

Muehlstein, L.K., Porter, D. & Short, F.T. (1991). *Labyrinthula zosterae* sp. nov., the causal agent of the wasting disease of eelgrass, *Zostera marina*. *Mycologia*, **83**, 180–91.

Neckles, H.A., Wetzel, R.L. & Orth, R.J. (1993). Relative effects of nutrient enrichment in grazing of epiphyton-macrophyte (*Zostera marina* L.) dynamics. *Oecologia*, **93**, 285–95.

Nielsen, S.L., Enríquez, S., Duarte, C.M. & Sand-Jensen, K. (1996). Scaling maximum growth rates across photosynthetic organisms. *Functional Ecology*, **10**, 167–175.

Nienhuis, P.H., De Bree, B.H.H., Herman, P.M.J., Holland, A.M.B., Verschuure, J.M. & Wessel, E.G.J. (1996). Twentyfive years of changes in the distribution and biomass of eelgrass, *Zostera marina*, in Grevelingen Lagoon, The Netherlands. *Netherlands Journal of Aquatic Ecology*, **30**, 107–17.

Nixon, S.W. (1995). Coastal marine eutrophication: a definition, social causes, and future concerns. *Ophelia*, **41**, 199–219.

Olesen, B. (1996). Regulation of light attenuation and eelgrass *Zostera marina* depth distribution in a Danish embayment. *Marine Ecology Progress Series*, **134**, 187–94.

Onuf, C.P. (1994). Seagrasses, dredging and light in Laguna Madre, Texas, U.S.A. *Estuarine, Coastal and Shelf Science*, **39**, 75–91.

Onuf, C.P. (1996). Seagrass responses to long-term light reduction by brown tide in upper Laguna Madre, Texas: distribution and biomass patterns. *Marine Ecology Progress Series*, **138**, 219–31.

Orth, R.J. & van Montfrans, J. (1984). Epiphyte-seagrass relationships with an emphasis on the role of micrograzing: a review. *Aquatic Botany*, **18**, 43–69.

Paling, E.I., Van Keulen, M. & Wheeler, K.D. (1998). *Seagrass rehabilitation in Owen Anchorage, Western Australia*, Annual Report Aug. 1997–June 1998. Marine and Freshwater Research Laboratory, Murdoch University, Report No. MAFRA 98/4.

Panayotou, T. (1993). The environment in Southeast Asia: problems and policies. *Environmental Science and Technology*, **27**, 2270–4.

Pergent, G., Mendez, S., Pergen-Martini, C. & Pasqualini, V. (1999). Preliminary data on the impact of fish farming facilities on *Posidonia oceanica* meadows in the Mediterranean. *Oceanologica Acta*, **22**, 95–107.

Philippart, C.J.M. (1994). Interactions between *Arenicola marina* and *Zostera noltii* on a tidal flat in the Wadden Sea. *Marine Ecology Progress Series*, **111**, 251–7.

Poiner, I.R., Walker, D.I. & Coles, R.G. (1989). Regional studies – Seagrasses of tropical Australia. In *Biology of Seagrasses*, ed. A.W.D Larkum, A.J. McComb & S.A. Shepherd, pp. 279–303. Amsterdam: Elsevier.

Porter, D. & Muehlstein, L.K (1989). A species of *Labyrinthula* is the prime suspect as the cause of a massive die off of the seagrass *Thalassia testudinum* in Florida Bay. *Mycological Society American Newsletter*, **40**, 43.

Preen, A.R., Lee Long, W.J. & Coles, R.G. (1995). Flood and cyclone related loss, and partial recovery, of more than 1000 km2 of seagrass in Hervey Bay, Queensland, Australia. *Aquatic Botany*, **52**, 3–17.

Ralph, P.J. (1998a). Photosynthetic response of laboratory-cultured *Halophila ovalis* to thermal stress. *Marine Ecology Progress Series*, **171**, 123–30.

Ralph, P.J. (1998b). Photosynthetic responses of *Halophila ovalis* (R.Br.) Hook. *f.* to osmotic stress. *Journal of Experimental Marine Biology and Ecology*, **227**, 203–20.

Ralph, P.J. & Burchett, M.D. (1995). Photosynthetic responses of the seagrass *Halophila ovalis* (R.Br.) Hook. f. to high irradiance stress, using chlorophyll *a* fluorescence. *Aquatic Botany*, **51**, 55–66.

Ralph, P.J. & Burchett, M.D. (1998). Impact of petrochemicals on the photosynthesis of *Halophila ovalis*, using chlorophyll fluorescence. *Marine Pollution Bulletin*, **36**, 429–36.

Robblee, M.B., Barber, T.R., Carlson, P.R., Jr., Durako, M.J., Fourqurean, J.W., Muehlstein, L.K.,

Porter, D., Yarbro, L.A., Zieman, R.T. & Zieman, J.C. (1991). Mass mortality of the tropical seagrass *Thalassia testudinum* in Florida Bay (USA). *Marine Ecology Progress Series*, **71**, 297–9.

Sand-Jensen, K. & Borum, J. (1991). Interactions among phytoplankton, periphyton, and macrophytes in temperate freshwaters and estuaries. *Aquatic Botany*, **41**, 137–76.

Shepherd, S.A., McComb, A.J., Bulthuis, D.A., Neverauskas, V., Steffensen, D.A. & West, R. (1989). Decline of seagrasses. In *Biology of Seagrasses*, ed. A.W.D. Larkum, A.J. McComb, & S.A. Shepherd, pp. 346–93. Amsterdam: Elsevier.

Short, F.T. & Burdick, D.M. (1996). Quantifying seagrass habitat loss in relation to housing development and nitrogen loading in Waquoit Bay, Massachusetts. *Estuaries*, **19**, 730–9.

Short, F.T. & Wyllie-Echeverria, S. (1996). Natural and human-induced disturbance of seagrasses. *Environmental Conservation*, **23**, 17–27.

Short, F.T. & Neckles, H.A. (1999). The effects of global climate change on seagrasses. *Aquatic Botany*, **63**, 169–196.

Short, F.T., Mathieson, A.C. & Nelson, J.I. (1986). Recurrence of the eelgrass wasting disease at the border of New Hampshire and Maine, USA. *Marine Ecology Progress Series*, **29**, 89–92.

S.A., Short, F.T., Burdick, D.M. & Kaldy, J.E. III (1995). Mesocosm experiments quantify the effects of eutrophication on eelgrass, *Zostera marina*. *Limnology & Oceanography*, **40**, 740–9.

Silberstein, K., Chiffings, A.W. & McComb, A.J. (1986). The loss of seagrass in Cockburn Sound, Western Australia. III. The effect of epiphytes on productivity of *Posidonia australis* Hook F. *Aquatic Botany*, **24**, 355–71.

Smith, S.V. (1981). Marine macrophytes as a global carbon sink. *Science*, **211**, 838–40.

Smith, R. & Baker, K. (1979). Penetration of UV-B and biologically effective dose-rates in natural waters. *Photochemistry and Photobiology*, **32**, 367–74.

Tack, J. & Polk, P. (1999). The influence of tropical catchments upon the coastal zone: modelling the links between groundwater and mangrove losses in Kenya, India/Bangladesh and Florida. In *The Sustainable Management of Tropical Catchments*, pp. 359–71, ed. D. Harper & T. Brown. John Wiley & Sons, Chichester.

Talbot, M.M.B., Knoop, W.T. & Bate, G.C. (1990). The dynamics of estuarine macrophytes in relation to flood/siltation cycles. *Botanica Marina*, **33**, 159–64.

Taylor, D.I., Nixon, S.W., Granger, S.L., Buckely, B.A., McMahon, J.P. & Lin, H.-J. (1995). Responses of coastal lagoon plant communities to different forms of nutrient enrichment – a mesocosm experiment. *Aquatic Botany*, **52**, 19–34.

Teramura, A.H. & Sullivan, J.H. (1994). Effects of UV-B radiation on photosynthesis and growth of terrestrial plants. *Photosynthetic Research*, **39**, 463–73.

Terrados, J., Duarte, C.M., Fortes, M.D., Borum, J., Agawin, N.S.R., Bach, S., Thampanya, U., Kamp-Nielsen, L., Kenworthy, W.J., Geertz-Hansen, O. & Vermaat, J. (1998). Changes in community structure and biomass of seagrass communities along gradients of siltation in SE Asia. *Estuarine, Coastal and Shelf Science*, **46**, 757–68.

Terrados, J., Duarte, C.M., Kamp-Nielsen, L., Agawin, N.S.R., Gacia, E., Lacap, D., Fortes, M.D., Borum, J., Lubanski, M. & Greve, T. (1999). Are seagrass growth and survival affected by reducing conditions in the sediment? *Aquatic Botany*, **65**, 175–98.

Tomasko, D.A. & Lapointe, B.E. (1991). Productivity and biomass of *Thalassia testudinum* as related to water column nutrient availability and epiphyte levels: field observations and experimental studies. *Marine Ecology Progress Series*, **75**, 9–17.

Tomasko, D.A., Dawes, C.A. & Hall, M.O. (1996). The effects of anthropogenic nutrient enrichment on turtle grass (*Thalassia testudinum*) in Sarasota Bay, Florida. *Estuaries*, **19**, 448–56.

Tréguer, P., Nelson, D.M., van Beusekom, A.J., DeMaster, D.J., Leynart, A. & Quéguiner, B. (1995). The silica balance in the world ocean: a reestimate. *Science*, **268**, 375–9.

Turner, N.C. & Bell, A.M. (1973). The ethnobotany of the Southern Kwakiutl Indians of British Columbia. *Economic Botany*, **27**, 257–310.

Van Katwijk, M.M., Vergeer, L.H.T., Schmitz, G.H.W. & Roelofs, J.G.M. (1997). Ammonium toxicity in eelgrass *Zostera marina*. *Marine Ecology Progress Series*, **157**, 159–73.

Van Katwijk, M.M., Schmitz, G.H.W., Hanssen, L.S.A.M.& Den Hartog, C. (1998). Suitability of *Zostera marina* populations for transplantation to the Wadden Sea as determined by a mesocosm shading experiment. *Aquatic Botany*, **60**, 283–305.

Van Katwijk, M.M., Schmitz, G.H.W., Gasseling, A.P. & Van Avesaath, P.H. (1999). The effects of salinity and nutrient load and their interaction on *Zostera marina* L. *Marine Ecology Progress Series*, **190**, 155–65.

Vergeer, L.H.T. & Den Hartog, C. (1991). Occurrence of wasting disease in *Zostera noltii*. *Aquatic Botany*, **40**, 155–63.

Vergeer, L.H.T. & Den Hartog, C. (1994). Omnipresence of Labyrinthulaceae in seagrasses. *Aquatic Botany*, **48**, 1–20.

Vermaat, J.E., Agawin, N.S.R., Fortes, M.D., Duarte, C.M., Marbà, Enríquez, S. & van Vierssen, W. (1997). The capacity of seagrasses to survive increased turbidity and siltation: the significance of growth form and light use. *Ambio*, **26**, 499–504.

Vidal, M., Duarte, C.M. & Sanchez, M.C. (1999) Coastal eutrophication research in Europe. Progression and imbalances. *Marine Pollution Bulletin*, **38**, 851–4.

Walker, D.I., Lukatelich, R.J., Bastyan, G. & McComb, A.J. (1989). Effect of boat moorings on seagrass beds near Perth, Western Australia. *Aquatic Botany*, **36**, 69–77.

Watson, R.A., Coles, R.G. & Lee Long, W.L. (1993). Simulation estimates of annual yield and landed value for commercial Penaeid prawns from a tropical seagrass habitat, Northern Queensland, Australia. *Australian Journal of Marine and Freshwater Research*, **44**, 211–19.

Wear, D.J., Sullivan, M.J., Moore, A.D. & Millie, D.F. (1999). Effects of water-column enrichment on the production dynamics of three seagrass species and their epiphytic algae. *Marine Ecology Progress Series*, **179**, 201–13.

Williams, S.L. & Davis, C.A. (1996). Population genetic analyses of transplanted eelgrass (*Zostera marina*) beds reveal reduced genetic diversity in southern California. *Restoration Ecology*, **4**, 163–80.

Williams, S.L. & Ruckelshaus, M.H. (1993). Effects of nitrogen availability and hervibory on eelgrass (*Zostera marina*) and epiphytes. *Ecology*, **74**, 904–18.

Wu, R.S.S. (1995). The environmental impact of marine fish culture: towards a sustainable future. *Marine Pollution Bulletin*: **31**, 159–66.

Zieman, J.C. & Wood, E.J.F. (1975). Effects of thermal pollution on tropical-type estuaries, with emphasis on Biscayne Bay, Florida. In *Tropical Marine Pollution*, ed. E.J.F.Wood & R.E. Johannes, Chapter 5. Amsterdam: Elsevier.

Zieman, J.C., Orth, R., Phillips, R.C., Thayer, G.W. & Thorhaug, A. (1984). The effects of oil on seagrass ecosystems. In *Restoration of Habitats Impacted by Oil Spills*, ed. J. Cairns & A. Buikema, pp. 37–64. Stoneham: Butterworth.

Zieman, J.C., Fourqurean, J.W. & Iverson, R.L. (1989). Distribution, abundance and productivity of seagrasses and macroalgae in Florida Bay. *Bulletin of Marine Science*, **44**, 292–311.

Zimmerman, R.C., Kohrs, D.G. & Alberte, R.S. (1996). Top-down impact through a bottom-up mechanism: the effect of limpet grazing on growth, productivity and carbon allocation of *Zostera marina* L. (eelgrass). *Oecologia*, **107**, 560–67.

Index

acclimatization, 106, 110
acetate, 158, 163
acetazolamide, 117
acetylene reduction method, 182, 186
adaptation, 82, 99, 105
age structure, 75
alcohols, 158, 163
allometric scaling, 42–7
ammonium, 19, 124, 125, 126, 127,
    128, 137, 161, 182, 238, 259
  adsorption, 125
amphipods, 156, 206, 217, 236
anaerobic metabolism, 38
anaerobiosis, 112, 113
anoxia, 112, 265
anoxic/anaerobic conditions, 111, 112,
    172, 173, 264
anoxic sediment, 5, 19, 112, 125, 158,
    160, 168, 183
anthers, 78
apatite, 125
apical dominance, 52, 55
aquaculture, 268
ATP, 112, 165
ATP-ase, 118

bacterioplankton, 177
bicarbonate/HCO_3^-, 116, 117, 118,
    119, 120, 121, 122, 274
biodiversity, 249; *see also* diversity
biogeochemical cycles, 23
biogeochemical processes, 92
biogeography, 6–9
biomass
  allocation, 58, 133

bacterial, 175, 176
belowground, 109, 110, 113, 155
community, 90
decline, 115
epiphyte, 236, 237, 261, 262, 265
fluctuations, 115
formation, 130
microbial, 175
of leaves, 91, 109, 235, 267
periphyton, 236
phytoplankton, 250, 261
plant, 109, 110, 232
production, 110, 124, 137, 173
root/rhizome, 183, 184
seagrass, 21, 73, 237, 250, 251, 260,
    267
bioturbation, 157, 234, 256
birds, 154, 155, 179
bivalves, 181, 224, 225, 226, 228, 238
blooms, 102, 261, 265
bracteae, 66
branching, 51–56, 58, 60, 73, 84
burial, 11, 17, 23, 59, 69, 70, 72, 78, 79,
    171–173, 251, 253, 266, 268
burrowing, 90, 158, 226, 238, 269

calcium content, 249
carbohydrates, 111, 112, 166, 174, 232,
    248, 249, 261
carbonate/CO_3^{2-}, 116, 117; *see also*
    carbonate sediments
carbon
  balance, 59, 108–110, 235, 261, 264,
    274
  content, 123, 124